Retrotransposons
and
Human Disease
L1 Retrotransposons as a
Source of Genetic Diversity

Retrotransposons
and
Human Disease
L1 Retrotransposons as a
Source of Genetic Diversity

editor

Abram Gabriel
Rutgers University, USA

World Scientific

NEW JERSEY · LONDON · SINGAPORE · BEIJING · SHANGHAI · HONG KONG · TAIPEI · CHENNAI · TOKYO

Published by

World Scientific Publishing Co. Pte. Ltd.
5 Toh Tuck Link, Singapore 596224
USA office: 27 Warren Street, Suite 401-402, Hackensack, NJ 07601
UK office: 57 Shelton Street, Covent Garden, London WC2H 9HE

Library of Congress Cataloging-in-Publication Data
Names: Gabriel, Abram, editor.
Title: Retrotransposons and human disease : L1 retrotransposons as a source of genetic diversity /
    editor, Abram Gabriel.
Description: Hackensack, New Jersey : World Scientific, [2023] |
    Includes bibliographical references and index.
Identifiers: LCCN 2022000538 | ISBN 9789811249211 (hardcover) |
    ISBN 9789811249228 (ebook for institutions) | ISBN 9789811249235 (ebook for individuals)
Subjects: MESH: Retroelements--genetics | Genetic Variation | RNA-Directed DNA Polymerase
Classification: LCC QH443 | NLM QU 470 | DDC 572.8/77--dc23/eng/20220201
LC record available at https://lccn.loc.gov/2022000538

British Library Cataloguing-in-Publication Data
A catalogue record for this book is available from the British Library.

Cover image:
Simplified representation of the life cycle of a retrotransposon
Courtesy of Dr. Marius Walter
Reproduced with the Creative Commons Attribution-Share Alike 4.0 International license

For any available supplementary material, please visit
https://www.worldscientific.com/worldscibooks/10.1142/12642#t=suppl

Typeset by Stallion Press
Email: enquiries@stallionpress.com

Preface

I was asked to edit this monograph in 2019, pre-COVID, by Christopher Davis. In 2010, I had a severe stroke while at my lab at Rutgers. I was fortunate to be able to continue teaching, but the stroke's effects forced me to give up my lab. This offer gave me the irresistible opportunity to continue my scientific learning.

As I began identifying and selecting contributors to this monograph, I could see that a whole new generation from around the world was now studying the many aspects of L1 biology. I spent my foreshortened career in science studying LTR retrotransposons while many of my colleagues also studied L1. I began to reach out to them, and found, fairly quickly, experts in the topics that this monograph covers.

Then, COVID hit, and all bets were off. Miraculously, the only chapter we lost because of COVID-19 was about L1 and aging. Some were long delayed, but eventually all the other authors came through. I was gratified by the quality of their work. The authors were a combination of senior scientists and an array of new investigators.

Reading through the chapters, I was struck by a few things. Many mentioned the presumed status of target primed reverse transcription (TPRT) with respect to human L1 movement and our lack of knowledge about subsequent steps in the mechanism of L1 replication. TPRT is a model from the 1980s and I was surprised by the lack of progress in this fundamental area. I was also struck by the emergence of both bioinformatics and gene knockouts over the last decade. No doubt this is due to advances in next-generation sequencing and

CRISPR technology. Finally, all these changes make me feel a bit antiquated. No matter how much you read, there's nothing like being in the lab, either doing experiments or writing papers. It can be slow, but once you prove something, it's true for all time, whether it's important or ultimately trivial.

This monograph contains nine chapters, all relating in some way to the relationship between L1 and human disease.

Chapter 1: Arkhipova and her colleagues underscored how common the reverse transcriptases (RTs), one of the enzymes found in L1, are to all three kingdoms of life, and how important RTs are to all organisms. For example, the enzyme telomerase, found in almost all eukaryotic species, has been shown to be a specialized RT.

Chapter 2: Miller and Le Grice describe the many dimensions of RTs of retrotransposons. For instance, they show that, unlike HIV-1, Ty3 RT is a homodimer where the DNA polymerase and RNase H activities are derived from separate subunits.

Chapter 3: The Han group writes about the importance of model systems as well as the fact that such systems can presage work on human L1, sometimes by decades. They also question certain details about the presumed roles of L1 in human biology. For example, they first make the distinction between germline transpositions and somatic events — which are not passed on to the next generation — and discuss the conjecture that L1 is related to infertility.

Chapter 4: Yang *et al.* tell the fascinating story of Alu movement in the human genome with its use of L1 ORF2. The often overshadowed Alu elements are an important source of human disease and their life cycles are much like those for L1.

Chapter 5: Kazazian updates his list from 2016 of monogenetic human disorders caused by L1. His long career in L1 dates back at least to 1988 when he and colleagues showed that two boys had haemophilia as the result of L1 insertions in their factor VIII genes. As readers are likely to know, Kazazian has continued his contributions to the field of human L1 biology ever since. Unfortunately, after writing that piece he died. His enthusiasm was infectious and he will be sorely missed.

Chapter 6: The Garcia-Perez group tells the engrossing tale of the role of L1s in early embryogenesis. They also make caveats to the assumption among some in the field that L1 retrotransposition is involved in the genesis of neurodevelopmental diseases such as schizophrenia or the Aicardi-Goutieres syndrome.

Chapter 7: Boeke and collaborators show the differences between non-LTR-containing elements, like human L1, and LTR-containing retrotransposons, like Ty1. They also discuss host factors that limit the retrotransposition of either type of element. Boeke was among those who invented the term retrotransposon. I had the honor of working in his lab at Johns Hopkins between 1988 and 1992. While he may have moved his lab to NYU, to me, he will always be the quintessential Hopkins scientist.

Chapter 8: The Gage group makes it clear that while the literature strongly suggests that the adult human brain is home to L1 mosaicism, the research still needs tightening up. For instance, most of the work has been done with mice. How humans respond is both more difficult and more relevant.

Chapter 9: Ardeljan's and Burns' piece on L1 and cancer is exhaustive, but not overstated. L1s are clearly very active in certain tumors. But, whether this is clinically relevant still awaits determination.

I want to thank the reviewers who I tried to keep anonymous and who read over the chapters for the accuracy of the references. They were Marlene Belfort, Stephane Boissinot, Joan Curcio, Geoff Faulkner, David Finnegan, Dave Garfinkel, Partho Ghosh, Steve Goff, and Molly Hammell. I also want to thank World Scientific Publishing for realizing this is an important area, especially Ms. Xiao Ling who is the desk editor of this volume. I also appreciate the secretarial help of Ms. Katie Hawn. And of course, there is my wife, Janet Heroux, who witnessed all my moods.

This is likely to be my last contribution to the field of retrotransposons. It has been a pleasure working with colleagues of long duration, and to encounter some of the emerging minds of the field.

All in all, this is where basic science meets clinical medicine. Enjoy reading it; I know I did.

Contents

**Chapter 3 Experimental Systems for the Study of
Non-LTR (LINE) Retrotransposons**   57

Ivana Celic and Jeffrey S. Han

Chapter 4 Alu Elements and Human Disease   77

*Hanlin Yang, Maria E. Morales, and
Astrid M. Roy-Engel*

Chapter 1
The Diversity of Reverse Transcriptases

Blair G. Paul*, Irina A. Yushenova*, and Irina R. Arkhipova*

1. Introduction

Reverse transcriptases (RTs), aka RNA-dependent DNA polymerases, are enzymes which can perform complementary DNA (cDNA) synthesis using RNA as a template. RTs occupy a very special place among polymerases, as reverse transcription is believed to have played a major role in the evolutionary transition from the primordial "RNA world" to the extant life forms based on DNA as the hereditary material.[1] In other words, the ability to polymerize deoxynucleotides in an RNA-templated fashion may well have been the centerpiece of the transition from an RNA-based genome to a DNA-based genome early in the history of life on Earth.

Template-dependent polymerization permits copying and transmission of the genetic code, written in the form of the primary nucleotide sequence, to subsequent generations, as well as translation of the code triplets into the sequence of amino acids in polypeptides. The apparent unidirectionality of this information flow from DNA to RNA to protein, called the Central Dogma of molecular biology,[2] was challenged by Temin and Baltimore in 1970, when they discovered RTs in retroviruses (then called RNA tumor viruses).[3,4] Subsequently,

*Marine Biological Laboratory, Woods Hole, MA 02543, USA

RTs became critical targets in studies of retroviral diseases such as HIV/AIDS.[5] Furthermore, the ability of RTs to synthesize DNA using RNA templates made them indispensable molecular tools for cDNA synthesis and RT-PCR. These cornerstone methodologies, which are used in basic science, biotechnology, and biomedicine, were recently supplemented by RT-Cas9 prime genome editing.[6] Nevertheless, RTs remain the most enigmatic enzymes among polymerases, even though recognition of their diversity in the past 50 years has increased dramatically, from reverse-transcribing viruses to a multitude of genetic elements found in every kingdom and domain of life.

RTs from different groups differ by the degree of phylogenetic relatedness, by overall structural organization of the RT-bearing genetic elements, and by arrangement of various N- and C-terminal domains fused to the common central core responsible for polymerization. The core comprises the seven highly conserved catalytically important motifs #1–7[7] and adopts the characteristic "right-hand" fold consisting of fingers, palm, and thumb,[8] also seen in certain other polymerases (e.g., T7 pol and Klenow fragment of DNA pol I).[7,9] In contrast to DNA polymerases, RTs lack proofreading activity, resulting in much higher error rates.

Several major groups of RTs, described in detail below, can be distinguished. Bacterial and archaeal RTs are mostly represented by group II introns (G2I), retrons, retroplasmids, diversity-generating retroelements (DGRs), RTs associated with CRISPR-Cas loci or abortive phage infection (Abi), and "orphan" RTs. Eukaryotic RTs include those found in LTR-retrotransposons/retroviruses (e.g., yeast Ty3), non-LTR retrotransposons (LINEs, e.g., mammalian L1), Penelope-like elements (PLEs), and catalytic subunits of telomerases (TERTs). Two viral families using reverse transcription in their life cycle, hepadnaviruses and caulimoviruses (collectively called pararetroviruses),[10] harbor RTs that are closely related to LTR-retrotransposon RTs and are believed to have a hybrid origin resulting from RT capture by a DNA virus. Finally, RNA-dependent RNA polymerases (RdRPs) from ssRNA and dsRNA viruses, extensively studied at the structural and functional level, are distantly related to RTs.[11,12] High-resolution 3D structures have been obtained for retroviral RTs, TERTs, yeast Ty3, and bacterial group II intron RTs,[13–19] but are still unavailable for most RT types, including LINEs. Here, we aim to survey the diversity of RTs from prokaryotes to eukaryotes, focusing on the role of RTs in retrotransposition and other cellular processes, their evolutionary histories, and their diverse functional characteristics, which may result in damage or benefit to the host cell.

2. Retroelements in Bacteria, Archaea, and their Viruses

Retroelements are widespread across most bacterial lineages, including bacteriophage, and they occur to a lesser extent in archaea and archaeal viruses. Prokaryotic retroelements are typically found in a few copies per genome and appear to be less proliferative compared with eukaryotic retroelements.[20] Nonetheless, bacterial and archaeal retroelements are phylogenetically and functionally diverse.[21,22] These modern retroelements most likely originated from ancient proliferative progenitors that have since evolved into the distinct extant groups that favor retrohoming, or targeted retromobility, over stochastic retrotransposition. Some classes of retroelements appear to be domesticated, offering beneficial utility to the host organism, wherein their regulation may be under tight control. To date, four retroelement classes have been functionally characterized in prokaryotes, including group II introns, retrons, DGRs, and abortive infection elements (Abi-like). Additionally, several groups of unclassified retroelements have been recently uncovered in genomic sequences, which appear to be phylogenetically distinct from most characterized retroelements.[20,23,24]

2.1. Group II introns

Catalytic RNA introns found in bacteria, archaea, and eukaryotic organelles are considered the ancestral elements from which spliceosomal introns evolved.[25-27] Group II introns (G2I) were first discovered in mitochondria and shortly thereafter they were identified in genomes from an array of bacterial lineages.[25] The genomic components of bacterial G2I are contained within a ~2–3 kbp locus (Fig. 1(A)), with 5′ and 3′ exons flanking intronic RNA. The intron sequence is, in turn,

(A) **Group-II-Introns**

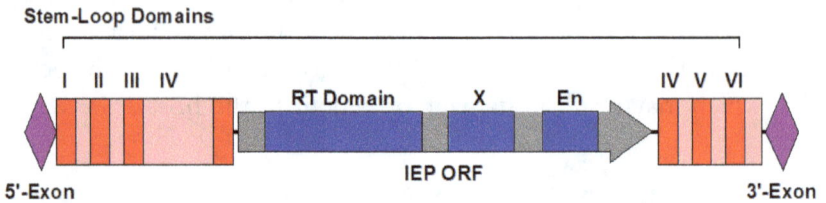

(B) **Retrons**

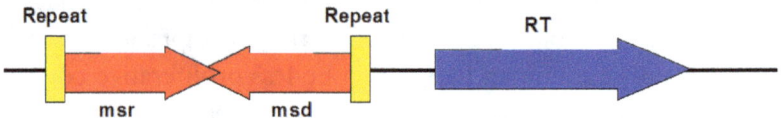

(C) **Diversity-Generating Retroelements**

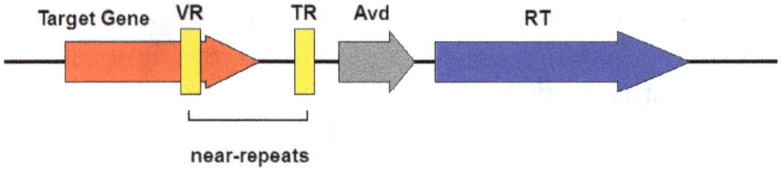

(D) **Abortive Infection System, AbiK**

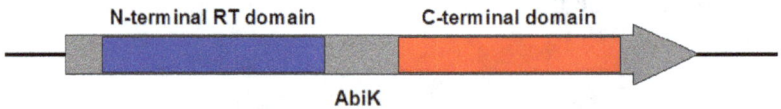

Figure 1. Genomic architectures for characterized bacterial and archaeal retroelements (not to scale). (A) Intronic stem-loop domains are shown in dark red, flanking the intron-encoded protein ORF (IEP; individual domains in blue). 5′ and 3′ flanking exons are indicated in purple. (B) Retron-encoded multicopy single-stranded DNA/RNA genes and flanking repeats are shown in red and yellow, respectively. (C) DGR cassette, with near repeats that differ at template adenines (VR, variable repeat; TR, template repeat), highlighted in yellow. A small grey ORF represents the accessory variability determinant (Avd) in the prototypical Bordetella phage DGR. (D) N-terminal RT and C-terminal conserved domain encoded by the AbiK ORF.

interrupted by a large internal intron-encoded protein gene (IEP), which is also referred to as a maturase for its role in RNA splicing.[28,29] Despite a variable primary sequence, the intron RNA has a conserved secondary structure, wherein several stem-loop domains (DI–DVI) branch from a large central loop. The first domain forms the primary structural scaffold of the ribozyme, which encodes exon-binding sites, EBS1 & EBS2.[30] The IEP ORF is encoded within the loop of domain IV and consists of a RT domain, a conserved domain X that contributes a thumb-like maturase component of RT, and an endonuclease-like En domain.[28]

Although G2Is require an IEP-RNA complex for splicing *in vivo*, their self-splicing activity has been demonstrated *in vitro* for both eukaryotic and bacterial introns.[25,31,32] To initiate self-splicing, the 2′-OH of a conserved adenosine in domain VI attacks the 5′ splice site to form a branched RNA lariat (Fig. 2(A)). Under normal cellular conditions, the IEP binds to the intron-containing transcript before splicing can proceed, to form a ribonucleoprotein (RNP) complex. After excision, the RNP mediates the integration of the intron at target DNA sites via reverse splicing (reviewed in Ref. [28]). Moreover, the IEP endonuclease domain assists with integration of newly synthesized cDNA into the chromosome. Chromosomal integration is enabled by DNA target-primed reverse transcription (TPRT), followed by DNA repair.[33,34] These elements are often referred to as selfish elements, in that G2I may replicate and proliferate through the bacterial chromosome in both specific and non-specific modes of retrohoming vs. retrotransposition, respectively.[34,35]

Group II introns are broadly distributed in bacteria and are less common in archaea, while in eukaryotes they are found in mitochondrial and chloroplast genomes.[20,28,36] This class of retroelement is a feature of bacterial chromosomes and plasmids, whereas G2I have not been identified in bacteriophage or archaeal viruses. Genome surveys and phylogenetic reconstruction highlight an apparent role of horizontal exchange and proliferation of G2I among members of related taxa.[37] Intriguingly, endosymbiotic bacteria possess an abundance of diverse G2I, suggesting genomic invasion and a complex history of horizontal transmission in these organisms.[38] In addition to IEP

(A) **Group-II-Introns**

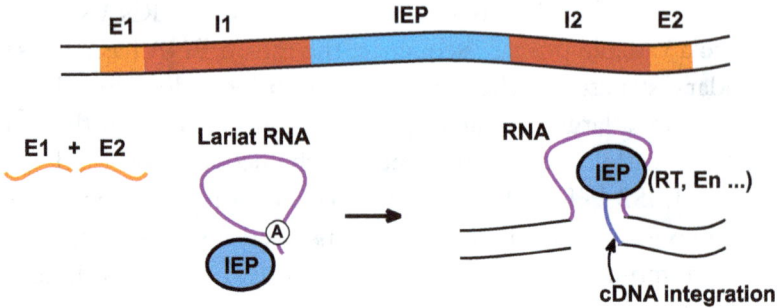

(B) **Retrons**

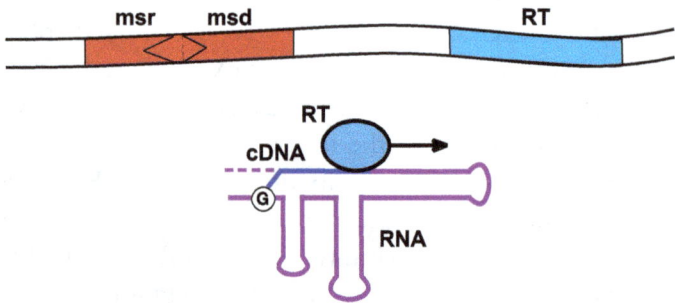

(C) **Diversity-Generating Retroelements**

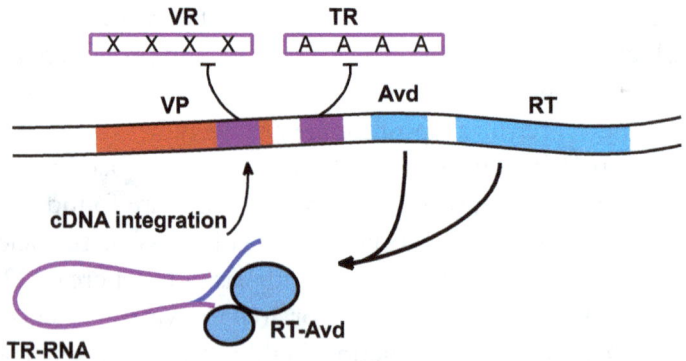

Figure 2.   Mechanistic overview for well-characterized bacterial retroelements. (A) RT-mediated splicing results in branched intronic ssRNA and excised, ligated exons. The branchpoint adenosine is circled on the RNA lariat. Intron mobility and integration via target primed reverse transcription are depicted, with subsequent DNA repair mechanisms not shown. (B) The essential, two-component retron architecture is

ORFs that have been phylogenetically characterized as G2I, a larger set of distantly related sequences constitutes several clades of G2I-like (G2L) retroelements, whose function(s) are yet undetermined.[20,39]

2.2. *Retrons*

Retrons were first discovered as small satellite DNA molecules that occur at a high copy number in bacterial cells.[40-43] The genomic features of a retron consist of multicopy single-stranded DNA (msDNA or msd) and overlapping msdRNA (or msr) adjacent to a RT gene (Fig. 1(B)). In addition, the msDNA/msdRNA region is typically flanked by two short complementary repeats.[44] Beyond carrying the essential RT gene, some retrons encode associated proteins that may be required for msDNA production. Retron RNA sequence and structure can vary across different hosts, where msdRNA can form several stem loops and the reverse-transcribed msDNA ranges from 48 to 163 bases in length.[44]

Bacterial retrons carry out partial reverse transcription of structural non-coding RNA.[42,45] A primary transcript is expressed that encodes, from 5′ to 3′, msd, msr, and RT. The retron RNA self-anneals and forms a hairpin secondary structure, with up to three stem-loops.[44] RT priming occurs at the msr-msd stem and DNA synthesis initiates at a guanosine residue branchpoint using msdRNA as template (Fig. 2(B)), followed by template degradation by host ribonuclease H.[45] Newly synthesized msDNA forms a heteroduplex with msdRNA that joins at a 2′-5′ phosphodiester bond. Importantly, retron activity was demonstrated *in trans* with msDNA production requiring only msd-msr and RT genes.[43] Although the functional

←─────────────────────────────────────

Figure 2. (***Continued***) shown as a genomic locus, comprising multicopy single-stranded DNA/RNA regions and RT. The secondary structure of retron RNA is shown in purple, with branched cDNA production indicated in blue. Circled G, guanosine residue branchpoint. (C) A chromosomal DGR cassette is depicted. Expanded boxes show the variable region (VR) and template region (TR) to highlight template adenines and corresponding mutable positions in VR. Avd, accessory variability determinant. Self-priming template RNA is shown in purple, with newly synthesized cDNA shown at a branched junction. A specific mechanism for the cDNA integration step is not depicted.

significance of the RNA–DNA complex remains unclear, recent studies have demonstrated that retrons may function as a novel toxin/antitoxin system that enables phage resistance.[46–48] Several retrons found in *Escherichia coli*, *Salmonella enterica*, and *Vibrio cholerae* appear to offer protection from phage lysis, which may be dependent on various effector proteins that mediate programmed cell death upon infection.[47] The genomic components of retrons are diverse in terms of functional domains and architecture.[49] For example, different retrons are dependent on unique functional domains in their effector proteins, including ribosyltransferases, nitrilases, and DNA-binding domains. Moreover, one retron can encode multiple RT genes, which are together required for antiphage activity.[48] The diversity of these tripartite retroelements raises the question of whether retrons serve functional roles in response to a broad range of cell conflicts, beyond phage lysis.

Retrons were initially discovered in Myxobacteria and characterized in a few other bacteria, but have more recently been identified broadly in many bacterial genomes, wherein they are encoded chromosomally, within prophage elements, and on plasmids.[20,21,41,45,50] Putative retrons have also been identified in the phylum Euryarchaeota,[51] but more broadly, this class of retroelements is likely a rare feature in archaeal genomes.[20] Moreover, bioinformatic surveys have uncovered >100 retron-like sequences from bacterial genomes, where these phylogenetically similar elements share a conserved region of the RT domain 7 that is associated with msRNA recognition.[20,24,51]

2.3. *Diversity-generating retroelements*

A remarkable class of domesticated retroelements can induce localized hypervariation in target genes of bacteria, archaea, and viruses.[52–54] DGRs were first discovered in *Bordetella* bacteriophages that undergo genetic variation to alternate host tropism between virulent and avirulent phases of *Bordetella*.[55,56] In the well-characterized phage system, an RT-dependent process involves transfer of information from a template sequence (TR) to a DNA target site (VR), wherein selective mutations in VR correspond to adenine positions in TR.[57] Importantly, the variant DNA is directed to the target gene only, while the template-encoding locus remains unchanged, thus permitting iterative rounds of diversification from a common starting sequence. The specific organization of these elements can vary in terms of synteny and number of targets, although a common architecture in most DGRs comprises the essential components encoded in a single cassette (Fig. 1(C)).

DGRs facilitate the mutation of specific genes through a form of retrohoming (Fig. 2(C)) that involves error-prone reverse transcription and targeted integration (reviewed in[52]). Diversification occurs through reverse transcription of the non-coding TR-RNA, which is homologous to the VR integration site (~100–200 bp), typically found near the 3′ end of the target gene(s). Following ncRNA expression, cDNA synthesis is initiated through template-primed reverse transcription.[58,59] While cDNA integration — the final step of retrohoming — likely requires host factors that await discovery, cDNA synthesis has been demonstrated *in vitro* with only DGR-RT, an accessory protein (Avd), ncRNA, and dNTPs.[59]

Although DGRs were initially described in bacteriophage and have since been identified in other viral sequences,[56] they may occur more frequently as chromosomal features in certain bacterial and archaeal lineages.[53,54,60,61] In general, DGRs are widespread across microbial taxa that occupy diverse environments, but they are most prevalent in genomes of uncultivated organisms from natural groundwater ecosystems.[53,61] DGRs appear to predominantly diversify genes that function in ligand-binding and host attachment,[62] but a vast majority of their variable proteins are functionally uncharacterized.

2.4. *Abortive infection systems: AbiA, AbiK, and Abi-P2*

A unique class of bacterial retroelements associated with abortive infection (Abi) consists of a RT-like gene, which is required for blocking phage replication, resulting in programmed cell death,[63,64] or phage exclusion.[65] Two of these RT-containing systems, AbiA and AbiK, are encoded on plasmids of the fermentative bacterium *Lactobacillus lactis*.[66] These retroelements are only found in a handful of bacterial genomes, all belonging to the class Bacilli.[20] Additionally, a distinct Abi-like retroelement, Abi-P2, occurs in the genomes of P2-like coliphage, where a RT domain is required for host exclusion of P5-like phage.[65]

Details are currently lacking on the mechanism underlying phage resistance by Abi-like retroelements. The sequence of AbiK is approximately 600 amino acids in length, comprising an N-terminal RT domain and an uncharacterized C-terminal domain (Fig. 1(D)). The RT domain in AbiK and other Abi-like retroelements has a putative active site with a conserved nucleotide-binding domain that has a diagnostic YXDD motif. However, AbiK RT lacks motifs of other domains that are typically present in bacterial retroelements and performs non-templated polymerization *in vitro*.[21,64] Moreover, although the C-terminal region of AbiK does not have a recognizable conserved domain, it is essential for phage resistance[63] and is hypothesized to play a role in protein priming.[64]

2.5. *Other uncharacterized retroelements*

Several clades of unclassified bacterial retroelements have been uncovered in bioinformatic surveys and most of their sequences are referred to as G2I-like (G2L), or simply "unknown".[22] Moreover, up to 17 distinct phylogenetic groups of bacterial and archaeal RTs have been defined, underscoring a complex evolutionary history and potentially vast, yet unknown functional diversity for retroelements in the two domains.[21] It remains unclear whether any of the unclassified retroelements serve a specific function to their host organisms or viruses, or whether they have diverged as selfish proliferative elements. One group of enigmatic retroelements is phylogenetically related to bacterial G2I and retrons, while occurring exclusively in small mitochondrial plasmids of fungal eukaryotes.[67-70] These mitochondrial retroplasmids, or mRPs, are able to integrate into mitochondrial DNA via either *de novo* initiation or 3′ template priming and reverse transcription.[71] Still, relatively little is known about the specific functional role of retroplasmids in fungal mitochondria.

Finally, a subset of uncharacterized bacterial retroelements is affiliated with natural CRISPR systems,[20,22,72] which provide bacterial and archaeal immunity against invading DNA or RNA.[73] Certain CRISPR cassettes have an RT ORF encoded alongside CRISPR-associated (Cas) genes, while in other instances, the RT domain appears grafted to the Cas domain.[24,74] More specifically, in some type-III CRISPR systems, a fused RT-Cas1 enzyme together with Cas2 and Cas6 enables recognition of invading RNA and integration of new spacer sequences into the repeat array for adaptive immunity.[75-77] Direct acquisition of the RNA spacer sequence involves CRISPR-associated ribonuclease, integrase, and RT properties, but the process likely depends on additional host factors. CRISPR-Cas-associated RTs appear to be widespread across bacterial phyla, but only a few archaeal representatives are found in the family *Methanosarcinaceae*.[72] A more representative taxonomic distribution of these elements could be uncovered through comprehensive metagenomic surveys. As with many other RTs identified through computational surveys, a vast majority of these putative CRISPR-RT systems remain experimentally uncharacterized.

RTs of most prokaryotic retroelements are not well adapted to the mobile element lifestyle which involves continuous relocation between different chromosomal sites — if they do change genomic locations, it may occur independently of a retroelement-specific mechanism. Group II introns represent a notable exception, as they can switch between the retrohoming and retromobility modes, and hence increase their genomic copy numbers. Furthermore, prokaryotic RTs display an amazing diversity of priming mechanisms: from target priming in G2I, to 2′-OH branching in retrons and possibly DGRs, to protein priming in AbiK, and *de novo* initiation in some retroplasmids. Association with an endonuclease allows G2I to operate more independently in heterologous environments, and the ability to target prime opens an avenue for immediate entry into new host targets. While no prokaryotic retroelements are capable of transposing in eukaryotic genomes, it is widely believed that the presence of G2I in organellar genomes indicates that G2I may have given rise to eukaryotic retroelements and to spliceosomal introns in the course of symbiogenesis and subsequent gene transfer from organellar endosymbionts into the nuclear genome of the ancestral eukaryote.[27,51]

3. Eukaryotic Reverse Transcriptases

In eukaryotes, the presence of the RT domain is usually recognized as the hallmark of a selfish mobile genetic element, such as a retrotransposon or a retrovirus. The degree of retrotransposon proliferation in eukaryotic genomes may vary from negligible to rampant, sometimes resulting in near takeover (up to 85%) of genomic DNA,[78,79] despite the existence of various genome defense mechanisms aimed at preventing uncontrolled expansion of mobile elements. Occasionally, however, retroelements benefit the host, or even undergo full domestication and become single-copy host genes, as discussed in the following section.

3.1. *Domesticated RTs: Telomerase, Prp8, and RVT genes*

In eukaryotes, recruitment of an RT to perform a host function is an exceptionally rare evolutionary event, possibly due to deleterious effects of reverse transcription acting on undesirable RNA templates. Nevertheless, one of the most important cellular enzymes responsible for maintenance of linear chromosome ends in eukaryotes, **telomerase reverse transcriptase (TERT)**, is a specialized RT believed to have originated from retrotransposons during early eukaryogenesis.[80,81] Indeed, TERTs are evolutionarily related to RTs of PLEs, having shared a common ancestor which apparently predates the origin of extant eukaryotes.[82-84] Telomerases are highly specialized RTs which maintain eukaryotic telomeres by addition of short G-rich repeated DNA sequences that are copied multiple times via reverse transcription of a specific region in the associated RNA template.[85] TERT represents the best-known example of a fully domesticated RT, i.e., it is encoded by a single-copy gene, which is not mobile and is unlinked from its specialized RNA template, the telomerase RNA (TER) encoded elsewhere in the genome. Domestication of telomerase at the dawn of eukaryotic evolution apparently involved substantial RT remodeling via additional domain acquisition. These domains served to adapt its catalytic properties to utilizing a short C-rich RNA template embedded in TER for extension of the exposed G-rich overhangs in chromosomal DNA, and to target RT to the chromosome termini in a complex with other proteins.[86,87]

Pre-mRNA-processing factor 8 (Prp8), the essential core component of the eukaryotic spliceosome, which regulates its assembly and conformation during pre-mRNA splicing, represents another landmark evolutionary co-option of the RT domain.[88-90] However, this RT-derived domain abandoned its catalytic function due to loss of the critical aspartates involved in polymerization, while retaining the ability of the thumb domain to interact with U5 snRNA and undergoing fusion to additional domains to facilitate interaction with other components of the spliceosomal machinery. Interestingly, the thumb domain is structurally most closely related to that of group II introns, which are thought to be the evolutionary precursors of spliceosomal introns.[17,18]

Reverse transcriptase-related (*rvt*) genes are a distinct class of domesticated RTs with unusual properties. Most notably, this is the only RT type found in both prokaryotes (Chloroflexi, Cyanobacteria, Bacteroidetes, Planctomycetes) and eukaryotes (except for organellar G2I),[91,92] although it has not yet been found in archaea. While *rvt* genes are predominant in fungi, they are also sporadically found in plants (certain mosses), protists (amoebas), and select metazoans such as insects (class Collembola) and rotifers (class Bdelloidea). Such cross-domain presence, together with signs of origination early in life before the appearance of eukaryotic forms, may imply a biological function which is applicable to both prokaryotes and eukaryotes. These genes usually exist in single copy, but can form 2–3-member gene families, and their syntenic environment, which can be traced in each family, confirms their non-mobile nature.[91] They evolve under purifying selection and may contain introns in evolutionarily conserved positions, resembling TERTs in this respect. Like TERT, RVTs can also exhibit terminal transferase activity *in vitro*. The domain structure of these genes clearly distinguishes them from other RT-containing elements due to the presence of a conserved N-terminal coiled-coil motif responsible for multimerization, a large insertion loop between motifs 2 and 3 of the RT core, and a unique C-terminal domain possibly involved in protein priming.[92] The function of *rvt* genes has yet to be elucidated.

3.2. *Classification of eukaryotic retrotransposons*

Since the first attempt at classification of eukaryotic transposable elements (TEs) by Finnegan,[93] all elements containing the RT domain, which transpose through an RNA intermediate, have been assembled into Class I, as opposed to Class II, which consists of all DNA transposons. The commonly accepted classification of retrotransposons, notwithstanding some differences in exact groupings, is based on their modular organization, i.e., specific combination of protein domains.[94,95] Additionally, all TEs, including Class I and Class II, can be divided into two subgroups based on the presence or absence of the enzymatic domain responsible for their mobilization: autonomous elements (e.g., LINE, Tc/mariner) and non-autonomous elements (e.g., SINE, MITE). Thus, Class I non-autonomous elements can only propagate when they are mobilized *in trans* by RT from autonomous elements. RT is the only enzyme which can effectively catalyze reverse transcription and is therefore the principal enzymatic domain common to all autonomous retrotransposons. Compared to their counterparts in prokaryotes, RT sequences of eukaryotic retro-elements are relatively well conserved, despite the enormous diversity of their genomic organization and replication strategies.[84]

All retrotransposons (Class I) are historically subdivided into two large categories distinguished by the presence/absence of long terminal repeats (LTRs): **non-LTR retrotransposons** (often collectively designated as LINEs) and **LTR retrotransposons** (retrovirus-like) (Fig. 3(A)). Two additional subclasses of eukaryotic retrotransposons, tyrosine recombinase-encoding (**YR-retrotransposons**, comprising three groups called DIRS, Pat, and Ngaro)[96] and **Penelope-like elements (PLEs)**,[82] are distinguishable by their terminal structures and distinct phylogenetic placement of their RTs (Fig. 3(B)). In addition to the universal presence of the RT domain, retrotransposons employ various phosphotransferase/endonuclease (EN) domains to integrate cDNA into new genomic locations: integrase (IN), tyrosine recombinase (YR), or apurinic-apyrimidinic (AP), restriction enzyme-like

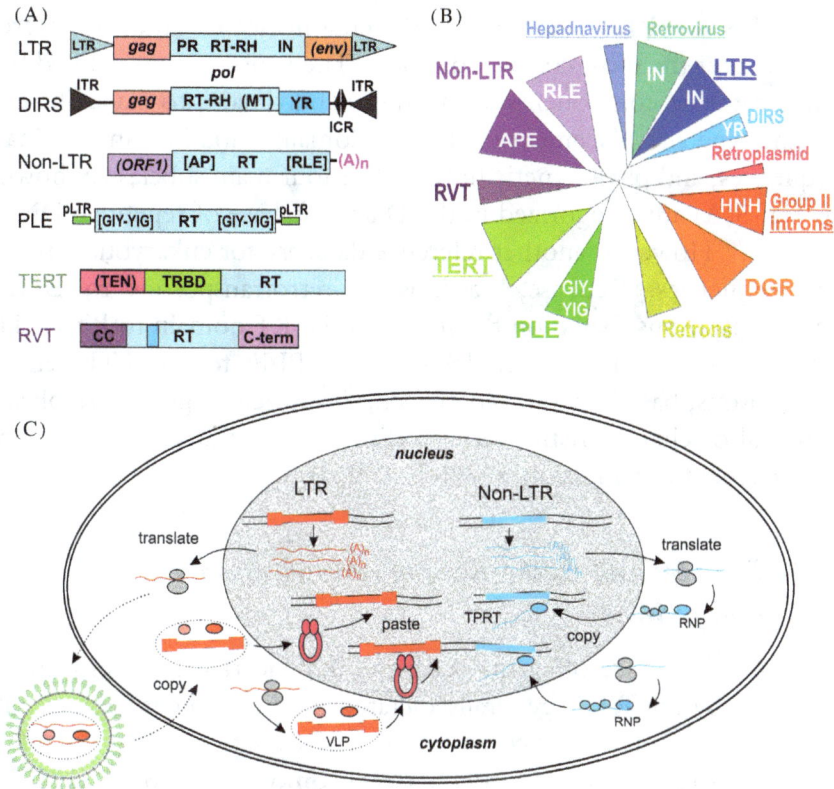

Figure 3. Properties of eukaryotic reverse transcriptases. (A) Structural features of major retrotransposon subclasses and domesticated RT genes. Shown are the functional modules of *gag, pol,* and *env* genes in different subclasses (RT, reverse transcriptase; RH, RNase H; IN, integrase; YR, tyrosine recombinase; GIY-YIG endonuclease; PR, protease; MT, methyltransferase; AP, apurinic-apyrimidinic endonuclease; RLE, restriction enzyme-like endonuclease); terminal structures (LTR, long terminal repeat; ITR, inverted terminal repeat; ICR, internal complementary region; pLTR, pseudo-LTR); and structural domains of domesticated RT genes (TEN, telomerase essential N-terminal domain; TRBD, telomerase RNA binding domain; CC, coiled-coil containing domain; C-term, C-terminal domain). Optional components are shown in parentheses. Square brackets indicate variable positioning. DGR, prokaryotic diversity-generating retroelements; PLE, Penelope-like elements; DIRS, a group of tyrosine recombinase-encoding LTR retrotransposons; TERT, telomerase reverse transcriptases. (B) Unrooted phylogenetic tree showing relationships between different RT types, including bacterial and eukaryotic retroelements and domesticated RT genes, with associated endonucleases (adapted from Ref. [91]). HNH, a type of endonuclease most frequently associated with group II introns. RT types with solved 3D structures are underlined. (C) Schematic replication cycles of LTR and non-LTR retrotransposons. TE-encoded proteins are shown by colored ovals; ribosomes, by gray ovals. VLP, virus-like particles; RNP, ribonucleoprotein particles; TPRT, target-primed reverse transcription; A$_n$, poly(A) tract in RNA. Retrotransposons encoding the *env* gene may have the potential to form extracellular particles, while those lacking *env* replicate intracellularly. Not to scale.

(REL-EN or RLE), and GIY-YIG endonucleases (the latter are named after the most conserved motif). The major groups defined by these catalytic components are further subdivided into clades based on the type and placement of the EN domain, variations in terminal sequences, and phylogenetic history. The four main subclasses shown in Fig. 3(A) are categorized in the Dfam database of repetitive DNA families.[97] However, another reference database for eukaryotic repetitive sequences, Repbase,[98] assigns YR-retrotransposons to LTR-retrotransposons, based on the fusion of the RT domain to RNase H (RH) in both LTR and YR elements, and PLEs to non-LTR retrotransposons, based on the variable length of their target site duplication, also characteristic for LINEs. For additional details on retroelement classification, see Refs. [99, 100].

3.3. *RT types and mechanisms of retrotransposon proliferation*

The exact sequence of events occurring in concert with reverse transcription by the RT is generally dictated by the element's structural organization and domain composition. There are two principal modes of priming reverse transcription: **extrachromosomally primed (EP)** and **target primed (TP)**.[101] The **EP mode** includes **LTR retrotransposons**, which were shown to utilize essentially the same replication mechanism as retroviruses (see the chapter by the Le Grice lab), as evidenced by similarities in structural organization of LTRs: conservation of enzymatically active domains including RT, RNase H (RH), protease (PR), and integrase (IN) with DDE catalytic residues common to most transposases; use of a host tRNA or self-priming to initiate reverse transcription; and the similar structure of replication intermediates.[102] Reverse transcription involves RT, RNA, and primer and takes place within a virus-like particle (VLP) formed by the product of the *gag* gene (Fig. 3(C)). This normally intracellular retrotransposition cycle may become extracellular if the retrotransposon encodes the optional *env* gene, the products of which can interact with the cell membrane to provide entry and egress. The double-stranded DNA product, which is synthesized by the RT within the

VLP in the cytoplasm through an elaborate scheme involving template jumps, is transported back into the nucleus, and integrated into chromosomal DNA by the integrase. Phylogenetic relatedness, and especially the presence of RT-RH domain fusions, implies that YR-retrotransposons follow essentially the same route, substituting IN with YR at the final integration step. Structural studies of Ty3 RT suggest that RT and RH activities reside on different subunits of the asymmetric RT homodimer formed upon substrate binding.[16]

The **TP mode** is apparently ancestral, as it is employed by more diverse retroelements. All known non-LTR retrotransposons use the mechanism of coupled reverse transcription/integration, named TPRT, whereby the EN cleaves the insertion site and the RT uses the exposed 3′-OH end of DNA to prime cDNA synthesis using the retroelement's transcript as a template.[101] Notably, the same mechanism is utilized by G2I and by the domesticated RT (TERT), which uses the naturally exposed 3′-OH groups that occur at chromosome ends to initiate TPRT. The TPRT mechanism was comprehensively described for the rDNA-specific R2 non-LTR retrotransposon of *Bombyx mori*,[103,104] elucidating the biochemical steps for the first-strand cDNA synthesis that is applicable to L1. A side effect of not requiring terminal repeats in the RNA template but requiring a poly(A) tract is that RTs of non-LTR retrotransposons are able to reverse transcribe in a single step not only RNAs of non-autonomous retroelements but also other cellular mRNA templates, generating processed pseudogenes. Highly efficient *trans*-mobilization of non-autonomous SINE elements, such as Alu, by the RT of autonomous LINEs results in high mutagenic potential and proliferative capacity of SINEs, which can outnumber the genomic share of L1 elements themselves.[105] LINE and SINE mobilization by the L1 RT is discussed in detail elsewhere in this book (see chapters by Boeke, Kazazian, Han, Roy-Engel). Unfortunately, structural studies of non-LTR retrotransposon RTs are still lagging, leaving LTR retrotransposons and TERTs as the only eukaryotic RTs investigated at the structural level.[15,16]

3.4. *RTs of non-LTR retrotransposons*

While phylogenetically close, non-LTR RTs are subdivided into two principal subclasses, based on the type and position of the EN domain within the pol polyprotein. The site-specific REL EN, located C-terminally to RT, typically confers site specificity to retrotransposon integration, recognizing the preferred motifs and positioning the RT domain at the nick to prime reverse transcription. Its similarity to archaeal Holliday junction resolvases may also help the RT initiate second-strand synthesis.[106] The rDNA-specific R2, spliced leader-specific NeSL of nematodes and CRE of trypanosomatids, or telomeric repeat-specific Gil/Genie retrotransposons of Giardia exemplify this group.[107–109] The AP-like EN (APE), located N-terminally to RT, exhibits less specificity in target recognition, making insertions of L1 and similar elements dispersed throughout the genome. However, in certain cases, APE may acquire insertion specificity, as happened in silkworm SART and TRAS retrotransposons, which insert into telomeric repeats added by telomerase.[110] Furthermore, in drosophilid insects, a complete evolutionary loss of telomerase and telomeric repeats coincided with accumulation of tandem head-to-tail arrays of specialized telomeric non-LTR retrotransposons HeT-A, TART, and TAHRE at chromosome ends.[111] Nevertheless, these TEs may not have achieved full domestication, as they are not found in some Drosophila species and could still exhibit signatures of evolutionary conflict.[112]

Even if full domestication is not achieved, a retrotransposon can cooperate with the host cell if the end result is advantageous for both. Telomeres represent an excellent genomic niche for such mutualistic interactions, whereby terminal and subterminal transposition of retro-elements, either working together with telomerase or in its absence, can compensate for loss of terminal DNA from chromosome ends during replication. It allows retrotransposons to proliferate in a genomic compartment with reduced potential for deleterious chromosome rearrangements and minimization of insertional damage to internally located host genes. Another such niche is the centromere, often consisting of highly repetitive satellite DNA blocks interrupted by islands of retroelements. In *Drosophila melanogaster*, non-LTR retrotransposons in these islands can provide binding sites for the centromeric histone CENP-A.[113] In maize, centromere-specific CRM LTR-retrotransposons help to localize the equivalent centromeric H3 histone CENH3 by restructuring chromatin loops via R-loop formation by circular RNAs.[114]

3.5. *RTs of Penelope-like elements*

PLEs are enigmatic yet the simplest eukaryotic retroelements, typically encoding a single large protein which in the originally discovered group of PLE represents a fusion of the RT domain with a GIY-YIG EN.[115,115a] EN-containing PLEs have been identified in animals, protists, and green algae/plants. Recombinant RT and EN from the prototype *Penelope* element of *Drosophila virilis* are catalytically active *in vitro*.[116] A special type of PLE that lacks the EN domain and is integrated in subtelomeric regions of chromosomes is found in diverse eukaryotes including protists, fungi, plants, and invertebrates.[83] Phylogenetically, PLEs cluster with TERT genes on the RT evolutionary tree[82] (Fig. 3(B)). Structurally, PLEs have direct or sometimes inverted repeats at the flanks and contain introns,[82] implying an unusual mechanism of retrotransposition, yet to be deciphered. In bdelloid rotifers, EN-deficient PLEs called *Athena* elements are organized in huge compound retroelements named *Terminons*, which combine the RT ORF and multiple other co-oriented ORFs, some of enzymatic and some of structural nature.[117] The 3′-ends of these giant telomeric retroelements contain a special hammerhead ribozyme RNA structure adjacent to reverse-complement telomeric repeats, which may facilitate attachment to the exposed G-rich overhangs of deprotected telomeres.[117] Such lineage-specific co-option relies on special properties of individual retrotransposons, in this case the propensity for terminal attachment. Indeed, this propensity may be acquired in more than one way: in drosophilids, the 3′-end of telomeric non-LTR elements at chromosome termini contains poly(A), while in the silkworm the specificity for telomeric repeats is determined by EN.

4. Concluding Remarks

Most eukaryotic RTs originate from retrotransposons or retroviruses and are thought to have no function other than to ensure proliferation of the selfish genetic elements carrying them.[118,119] Retrotransposons mostly cause damage when their insertion disrupts host genes or important regulatory elements. However, they may also disseminate their own promoters and enhancers throughout the genome, with the potential to diversify and modulate expression levels of nearby genes. RT domestication is exceptionally rare, and telomerases and *rvt* genes are so far the only known non-mobile catalytically active RTs. Despite the rarity of domestication of RT itself, recruitment of other retrotransposon components (*gag, env*) to perform functions beneficial to the host is a relatively common phenomenon.[120] Undeniably, RTs represent a potent evolutionary force which has been shaping and reshaping the genomes of prokaryotes and eukaryotes alike during millions of years of evolution and continues to do so in real-time.

Acknowledgments

We thank Marlene Belfort and Partho Ghosh for critical reading of the manuscript. Work in our laboratories is supported by the Gordon and Betty Moore and the G. Unger Vetlesen foundations (BP), the US National Science Foundation MCB-2139001 (IA, IY) and the US National Institutes of Health R01GM111917 (IA).

References

1. Gesteland RF, Cech TR, Atkins JF. The RNA World. 3rd ed. Cold Spring Harbor, NewYork: Cold Spring Harbor Laboratory Press; 2006.
2. Crick F. Central dogma of molecular biology. *Nature* 1970;227:561–3.
3. Baltimore D. RNA-dependent DNA polymerase in virions of RNA tumour viruses. *Nature* 1970;226:1209–11.
4. Temin HM, Mizutani S. RNA-dependent DNA polymerase in virions of Rous sarcoma virus. *Nature* 1970;226:1211–3.
5. Arts EJ, Hazuda DJ. HIV-1 antiretroviral drug therapy. *Cold Spring Harb Perspect Med* 2012;2:a007161.

6. Anzalone AV, Randolph PB, Davis JR, et al. Search-and-replace genome editing without double-strand breaks or donor DNA. *Nature* 2019;576:149–57.

7. Xiong Y, Eickbush TH. Origin and evolution of retroelements based upon their reverse transcriptase sequences. *EMBO J* 1990;9:3353–62.

8. Arnold E, Jacobo-Molina A, Nanni RG, et al. Structure of HIV-1 reverse transcriptase/DNA complex at 7 A resolution showing active site locations. *Nature* 1992;357:85–9.

9. Salgado PS, Koivunen MR, Makeyev EV, Bamford DH, Stuart DI, Grimes JM. The structure of an RNAi polymerase links RNA silencing and transcription. *PLoS Biol* 2006;4:e434.

10. Temin HM. Reverse transcription in the eukaryotic genome: retroviruses, pararetroviruses, retrotransposons, and retrotranscripts. *Mol Biol Evol* 1985;2:455–68.

11. Poch O, Sauvaget I, Delarue M, Tordo N. Identification of four conserved motifs among the RNA-dependent polymerase encoding elements. *EMBO J* 1989;8:3867–74.

12. Wolf YI, Kazlauskas D, Iranzo J, et al. Origins and evolution of the global RNA virome. *mBio* 2018;9.

13. Huang H, Chopra R, Verdine GL, Harrison SC. Structure of a covalently trapped catalytic complex of HIV-1 reverse transcriptase: implications for drug resistance. *Science* 1998;282:1669–75.

14. Das D, Georgiadis MM. The crystal structure of the monomeric reverse transcriptase from Moloney murine leukemia virus. *Structure* 2004;12:819–29.

15. Gillis AJ, Schuller AP, Skordalakes E. Structure of the Tribolium castaneum telomerase catalytic subunit TERT. *Nature* 2008;455:633–7.

16. Nowak E, Miller JT, Bona MK, et al. Ty3 reverse transcriptase complexed with an RNA-DNA hybrid shows structural and functional asymmetry. *Nat Struct Mol Biol* 2014;21:389–96.

17. Qu G, Kaushal PS, Wang J, et al. Structure of a group II intron in complex with its reverse transcriptase. *Nat Struct Mol Biol* 2016;23:549–57.

18. Zhao C, Pyle AM. Crystal structures of a group II intron maturase reveal a missing link in spliceosome evolution. *Nat Struct Mol Biol* 2016;23:558–65.

19. Stamos JL, Lentzsch AM, Lambowitz AM. Structure of a thermostable group II Intron reverse transcriptase with template-primer and its functional and evolutionary implications. *Mol Cell* 2017;68:926–39.

20. Simon DM, Zimmerly S. A diversity of uncharacterized reverse transcriptases in bacteria. *Nucleic Acids Res* 2008;36:7219–29.

21. Toro N, Nisa-Martinez R. Comprehensive phylogenetic analysis of bacterial reverse transcriptases. *PLoS One* 2014;9:e114083.

22. Zimmerly S, Wu L. An unexplored diversity of reverse transcriptases in bacteria. *Microbiol Spectr* 2015;3:MDNA3-0058-2014.

23. Kojima KK, Kanehisa M. Systematic survey for novel types of prokaryotic retroelements based on gene neighborhood and protein architecture. *Mol Biol Evol* 2008;25:1395–404.

24. Toro N, Martinez-Abarca F, Mestre MR, Gonzalez-Delgado A. Multiple origins of reverse transcriptases linked to CRISPR-Cas systems. *RNA Biol* 2019;16:1486–93.

25. Ferat JL, Michel F. Group II self-splicing introns in bacteria. *Nature* 1993;364:358–61.

26. Curcio MJ, Belfort M. Retrohoming: cDNA-mediated mobility of group II introns requires a catalytic RNA. *Cell* 1996;84:9–12.

27. Novikova O, Belfort M. Mobile group II Introns as ancestral eukaryotic elements. *Trends Genet* 2017;33:773–83.

28. Lambowitz AM, Zimmerly S. Mobile group II introns. *Annu Rev Genet* 2004;38:1–35.

29. Toor N, Keating KS, Taylor SD, Pyle AM. Crystal structure of a self-spliced group II intron. *Science* 2008;320:77–82.

30. Lambowitz AM, Belfort M. Introns as mobile genetic elements. *Annu Rev Biochem* 1993;62:587–622.

31. Peebles CL, Perlman PS, Mecklenburg KL, *et al.* A self-splicing RNA excises an intron lariat. *Cell* 1986;44:213–23.

32. van der Veen R, Arnberg AC, van der Horst G, Bonen L, Tabak HF, Grivell LA. Excised group II introns in yeast mitochondria are lariats and can be formed by self-splicing in vitro. *Cell* 1986;44:225–34.

33. Zimmerly S, Guo H, Perlman PS, Lambowitz AM. Group II intron mobility occurs by target DNA-primed reverse transcription. *Cell* 1995;82:545–54.

34. Cousineau B, Smith D, Lawrence-Cavanagh S, *et al.* Retrohoming of a bacterial group II intron: mobility via complete reverse splicing, independent of homologous DNA recombination. *Cell* 1998;94:451–62.

35. Cousineau B, Lawrence S, Smith D, Belfort M. Retrotransposition of a bacterial group II intron. *Nature* 2000;404:1018–21.

36. Toro N. Bacteria and Archaea Group II introns: additional mobile genetic elements in the environment. *Environ Microbiol* 2003;5:143–51.

37. Zimmerly S, Hausner G, Wu X-c. Phylogenetic relationships among group II intron ORFs. *Nucl Acids Res* 2001;29:1238–50.

38. Leclercq S, Giraud I, Cordaux R. Remarkable abundance and evolution of mobile group II introns in Wolbachia bacterial endosymbionts. *Mol Biol Evol* 2011;28:685–97.
39. Toro N, Martinez-Abarca F. Comprehensive phylogenetic analysis of bacterial group II intron-encoded ORFs lacking the DNA endonuclease domain reveals new varieties. *PLoS One* 2013;8:e55102.
40. Yee T, Furuichi T, Inouye S, Inouye M. Multicopy single-stranded DNA isolated from a gram-negative bacterium, *Myxococcus xanthus*. *Cell* 1984;38:203–9.
41. Lampson BC, Inouye M, Inouye S. Retrons, msDNA, and the bacterial genome. *Cytogenet Genome Res* 2005;110:491–9.
42. Inouye S, Hsu MY, Eagle S, Inouye M. Reverse transcriptase associated with the biosynthesis of the branched RNA-linked msDNA in Myxococcus xanthus. *Cell* 1989;56:709–17.
43. Lim D, Maas WK. Reverse transcriptase-dependent synthesis of a covalently linked, branched DNA-RNA compound in *E. coli* B. *Cell* 1989;56:891–904.
44. Simon AJ, Ellington AD, Finkelstein IJ. Retrons and their applications in genome engineering. *Nucleic Acids Res* 2019;47:11007–19.
45. Inouye M, Inouye S. msDNA and bacterial reverse transcriptase. *Annu Rev Microbiol* 1991;45:163–86.
46. Bobonis J, Mateus A, Pfalz B, *et al.* Bacterial retrons encode tripartite toxin/antitoxin systems. *bioRxiv* 2020:2020.06.22.160168.
47. Millman A, Bernheim A, Stokar-Avihail A, *et al.* Bacterial retrons function in anti-phage defense. *Cell* 2020;183:1551–61.
48. Gao L, Altae-Tran H, Böhning F, *et al.* Diverse enzymatic activities mediate antiviral immunity in prokaryotes. *Science* 2020;369:1077–84.
49. Mestre MR, González-Delgado A, Gutiérrez-Rus LI, Martínez-Abarca F, Toro N. Systematic prediction of genes functionally associated with bacterial retrons and classification of the encoded tripartite systems. *Nucleic Acids Res* 2020.
50. Rychlik I, Sebkova A, Gregorova D, Karpiskova R. Low-molecular-weight plasmid of *Salmonella enterica* serovar Enteritidis codes for retron reverse transcriptase and influences phage resistance. *J Bacteriol* 2001;183:2852–8.
51. Rest JS, Mindell DP. Retroids in archaea: phylogeny and lateral origins. *Mol Biol Evol* 2003;20:1134–42.
52. Guo H, Arambula D, Ghosh P, Miller JF. Diversity-generating retroelements in phage and bacterial genomes. *Microbiol Spectr* 2014;2: MDNA3-0029-2014.

53. Paul BG, Burstein D, Castelle CJ, *et al*. Retroelement-guided protein diversification abounds in vast lineages of Bacteria and Archaea. *Nat Microbiol* 2017;2:17045.

54. Wu L, Gingery M, Abebe M, *et al*. Diversity-generating retroelements: natural variation, classification and evolution inferred from a large-scale genomic survey. *Nucleic Acids Res* 2018;46:11–24.

55. Liu M, Deora R, Doulatov SR, *et al*. Reverse transcriptase-mediated tropism switching in Bordetella bacteriophage. *Science* 2002;295:2091–4.

56. Doulatov S, Hodes A, Dai L, *et al*. Tropism switching in Bordetella bacteriophage defines a family of diversity-generating retroelements. *Nature* 2004;431:476–81.

57. Handa S, Reyna A, Wiryaman T, Ghosh P. Determinants of adenine-mutagenesis in diversity-generating retroelements. *Nucleic Acids Res* 2021;49:1033–45.

58. Naorem SS, Han J, Wang S, *et al*. DGR mutagenic transposition occurs via hypermutagenic reverse transcription primed by nicked template RNA. *Proc Natl Acad Sci* U S A 2017;114:E10187–95.

59. Handa S, Jiang Y, Tao S, *et al*. Template-assisted synthesis of adenine-mutagenized cDNA by a retroelement protein complex. *Nucleic Acids Res* 2018;46:9711–25.

60. Vallota-Eastman A, Arrington EC, Meeken S, *et al*. Role of diversity-generating retroelements for regulatory pathway tuning in cyanobacteria. *BMC Genomics* 2020;21:664.

61. Roux S, Paul BG, Bagby SC, *et al*. Ecology and molecular targets of hypermutation in the global microbiome. *Nat Commun* 2021;12:3076.

62. Le Coq J, Ghosh P. Conservation of the C-type lectin fold for massive sequence variation in a Treponema diversity-generating retroelement. *Proc Natl Acad Sci* U S A 2011;108:14649–53.

63. Fortier LC, Bouchard JD, Moineau S. Expression and site-directed mutagenesis of the lactococcal abortive phage infection protein AbiK. *J Bacteriol* 2005;187:3721–30.

64. Wang C, Villion M, Semper C, Coros C, Moineau S, Zimmerly S. A reverse transcriptase-related protein mediates phage resistance and polymerizes untemplated DNA in vitro. *Nucleic Acids Res* 2011;39:7620–9.

65. Odegrip R, Nilsson AS, Haggård-Ljungquist E. Identification of a gene encoding a functional reverse transcriptase within a highly variable locus in the P2-like coliphages. *J Bacteriol* 2006;188:1643–7.

66. Chopin MC, Chopin A, Bidnenko E. Phage abortive infection in lactococci: variations on a theme. *Curr Opin Microbiol* 2005;8:473–9.

67. Kennell JC, Saville BJ, Mohr S, *et al.* The VS catalytic RNA replicates by reverse transcription as a satellite of a retroplasmid. *Genes Dev* 1995;9:294–303.

68. Chiang CC, Lambowitz AM. The Mauriceville retroplasmid reverse transcriptase initiates cDNA synthesis de novo at the 3′ end of tRNAs. *Mol Cell Biol* 1997;17:4526–35.

69. Simpson EB, Ross SL, Marchetti SE, Kennell JC. Relaxed primer specificity associated with reverse transcriptases encoded by the pFOXC retroplasmids of *Fusarium oxysporum*. *Eukaryot Cell* 2004;3:1589–600.

70. Galligan JT, Kennell JC. Retroplasmids: linear and circular plasmids that replicate via reverse transcription. In: Meinhardt F, Klassen R, editors. Microbial Linear Plasmids. Berlin, Heidelberg: Springer Berlin Heidelberg; 2007. p. 163–85.

71. Chen B, Lambowitz AM. De novo and DNA primer-mediated initiation of cDNA synthesis by the mauriceville retroplasmid reverse transcriptase involve recognition of a 3′ CCA sequence. *J Mol Biol* 1997;271:311–32.

72. Toro N, Martinez-Abarca F, Gonzalez-Delgado A. The reverse transcriptases associated with CRISPR-Cas systems. *Sci Rep* 2017;7:7089.

73. Brouns SJ, Jore MM, Lundgren M, *et al.* Small CRISPR RNAs guide antiviral defense in prokaryotes. *Science* 2008;321:960–4.

74. Koonin EV, Makarova KS, Zhang F. Diversity, classification and evolution of CRISPR-Cas systems. *Curr Opin Microbiol* 2017;37:67–78.

75. Silas S, Mohr G, Sidote DJ, *et al.* Direct CRISPR spacer acquisition from RNA by a natural reverse transcriptase-Cas1 fusion protein. *Science* 2016;351:aad4234.

76. Mohr G, Silas S, Stamos JL, *et al.* A reverse transcriptase-Cas1 fusion protein contains a Cas6 domain required for both CRISPR RNA biogenesis and RNA spacer acquisition. *Mol Cell* 2018;72:700–14.

77. González-Delgado A, Mestre MR, Martínez-Abarca F, Toro N. Spacer acquisition from RNA mediated by a natural reverse transcriptase-Cas1 fusion protein associated with a type III-D CRISPR-Cas system in *Vibrio vulnificus*. *Nucleic Acids Res* 2019;47:10202–11.

78. Rogers RL, Zhou L, Chu C, *et al.* Genomic takeover by transposable elements in the strawberry poison frog. *Mol Biol Evol* 2018;35:2913–27.

79. Jiao Y, Peluso P, Shi J, *et al.* Improved maize reference genome with single-molecule technologies. *Nature* 2017;546:524.

80.   Eickbush TH. Telomerase and retrotransposons: which came first? *Science* 1997;277:911–2.

81.   Nakamura TM, Cech TR. Reversing time: origin of telomerases. *Cell* 1998;92:587–90.

82.   Arkhipova IR, Pyatkov KI, Meselson M, Evgen'ev MB. Retroelements containing introns in diverse invertebrate taxa. *Nat Genet* 2003;33:123–4.

83.   Gladyshev EA, Arkhipova IR. Telomere-associated endonuclease-deficient Penelope-like retroelements in diverse eukaryotes. *Proc Natl Acad Sci* U S A 2007;104:9352–7.

84.   Koonin EV, Dolja VV. Virus world as an evolutionary network of viruses and capsidless selfish elements. *Microbiol Mol Biol Rev* 2014;78:278–303.

85.   Blackburn EH. Telomeres and telomerase: their mechanisms of action and the effects of altering their functions. *FEBS Lett* 2005;579:859–62.

86.   Arkhipova IR. Telomerase, retrotransposons, and evolution. In: Lue NF, Autexier C, editors. Telomerases: Chemistry, Biology, and Clinical Applications. Hoboken, NJ: John Wiley & Sons, Inc.; 2012. p. 265–99.

87.   Shay JW, Wright WE. Telomeres and telomerase: three decades of progress. *Nat Rev Genet* 2019;20:299–309.

88.   Galej WP, Oubridge C, Newman AJ, Nagai K. Crystal structure of Prp8 reveals active site cavity of the spliceosome. *Nature* 2013;493:638–43.

89.   Nguyen TH, Galej WP, Bai XC, et al. The architecture of the spliceosomal U4/U6.U5 tri-snRNP. *Nature* 2015;523:47–52.

90.   Wan R, Yan C, Bai R, et al. The 3.8 A structure of the U4/U6.U5 tri-snRNP: insights into spliceosome assembly and catalysis. *Science* 2016;351:466–75.

91.   Gladyshev EA, Arkhipova IR. A widespread class of reverse transcriptase-related cellular genes. *Proc Natl Acad Sci* U S A 2011; 108:20311–6.

92.   Yushenova IA, Arkhipova IR. Biochemical properties of bacterial reverse transcriptase-related (rvt) gene products: multimerization, protein priming, and nucleotide preference. *Curr Genet* 2018;64:1287–301.

93.   Finnegan DJ. Eukaryotic transposable elements and genome evolution. *Trends Genet* 1989;5:103–7.

94.   Wicker T, Sabot F, Hua-Van A, et al. A unified classification system for eukaryotic transposable elements. *Nat Rev Genet* 2007;8:973–82.

95. Kapitonov VV, Jurka J. A universal classification of eukaryotic transposable elements implemented in Repbase. *Nat Rev Genet* 2008;9:411–2.

96. Poulter RTM, Butler MI. Tyrosine recombinase retrotransposons and transposons. *Microbiol Spectr* 2015;3:MDNA3-0036-2014.

97. Storer J, Hubley R, Rosen J, Wheeler TJ, Smit AF. The Dfam community resource of transposable element families, sequence models, and genome annotations. *Mobile DNA* 2021;12:2.

98. Bao W, Kojima KK, Kohany O. Repbase update, a database of repetitive elements in eukaryotic genomes. *Mob DNA* 2015;6:11.

99. Arkhipova IR. Using bioinformatic and phylogenetic approaches to classify transposable elements and understand their complex evolutionary histories. *Mob DNA* 2017;8:19.

100. Kojima KK. Structural and sequence diversity of eukaryotic transposable elements. *Genes Genet Syst* 2020;94:233–52.

101. Beauregard A, Curcio MJ, Belfort M. The take and give between retrotransposable elements and their hosts. *Annu Rev Genet* 2008;42:587–617.

102. Arkhipova IR, Mazo AM, Cherkasova VA, Gorelova TV, Schuppe NG, Ilyin YV. The steps of reverse transcription of Drosophila mobile genetic elements and U3-R-U5 structure of their LTRs. *Cell* 1986;44:555–63.

103. Luan DD, Korman MH, Jakubczak JL, Eickbush TH. Reverse transcription of R2Bm RNA is primed by a nick at the chromosomal target site: a mechanism for non-LTR retrotransposition. *Cell* 1993;72:595–605.

104. Christensen SM, Ye J, Eickbush TH. RNA from the 5′ end of the R2 retrotransposon controls R2 protein binding to and cleavage of its DNA target site. *Proc Natl Acad Sci U S A* 2006;103:17602–7.

105. Dewannieux M, Heidmann T. LINEs, SINEs and processed pseudogenes: parasitic strategies for genome modeling. *Cytogenet Genome Res* 2005;110:35–48.

106. Khadgi BB, Govindaraju A, Christensen SM. Completion of LINE integration involves an open '4-way' branched DNA intermediate. *Nucleic Acids Res* 2019;47:8708–19.

107. Eickbush TH. R2 and Related Site-Specific Non-Long Terminal Repeat Retrotransposons. Washington, DC: ASM Press; 2002.

108. Gabriel A, Boeke JD. Reverse transcriptase encoded by a retrotransposon from the trypanosomatid *Crithidia fasciculata*. *Proc Natl Acad Sci U S A* 1991;88:9794–8.

109.   Arkhipova IR, Morrison HG. Three retrotransposon families in the genome of *Giardia lamblia*: two telomeric, one dead. *Proc Natl Acad Sci U S A* 2001;98:14497–502.

110.   Nichuguti N, Fujiwara H. Essential factors involved in the precise targeting and insertion of telomere-specific non-LTR retrotransposon, SART1Bm. *Sci Rep* 2020;10:8963.

111.   Pardue ML, DeBaryshe PG. Retrotransposons that maintain chromosome ends. *Proc Natl Acad Sci U S A* 2011;108:20317–24.

112.   Saint-Leandre B, Nguyen SC, Levine MT. Diversification and collapse of a telomere elongation mechanism. *Genome Res* 2019;29:920–31.

113.   Chang C-H, Chavan A, Palladino J, *et al.* Islands of retroelements are major components of Drosophila centromeres. *PLoS Biol* 2019;17:e3000241.

114.   Liu Y, Su H, Zhang J, Liu Y, Feng C, Han F. Back-spliced RNA from retrotransposon binds to centromere and regulates centromeric chromatin loops in maize. *PLoS Biol* 2020;18:e3000582.

115.   Evgen'ev MB, Arkhipova IR. Penelope-like elements – a new class of retroelements: distribution, function and possible evolutionary significance. *Cytogenet Genome Res* 2005;110:510–21.

115a. Craig RJ, Yushenova IA, Rodriguez F, Arkhipova IR. An ancient clade of Penelope-Like retroelements with permuted domains is present in the green lineage and protists, and dominates many invertebrate genomes. Mol Biol Evol. 2021; 38:5005-20.

116.   Pyatkov KI, Arkhipova IR, Malkova NV, Finnegan DJ, Evgen'ev MB. Reverse transcriptase and endonuclease activities encoded by Penelope-like retroelements. *Proc Natl Acad Sci U S A* 2004;101:14719–24.

117.   Arkhipova IR, Yushenova IA, Rodriguez F. Giant reverse transcriptase-encoding transposable elements at telomeres. *Mol Biol Evol* 2017;34:2245–57.

118.   Agren JA, Clark AG. Selfish genetic elements. *PLoS Genet* 2018;14:e1007700.

119.   Arkhipova IR. Neutral theory, transposable elements, and eukaryotic genome evolution. *Mol Biol Evol* 2018;35:1332–7.

120.   Cosby RL, Chang N-C, Feschotte C. Host–transposon interactions: conflict, cooperation, and cooption. *Genes Dev* 2019;33:1098–116.

Chapter 2
Ty3 and Related LTR-Retrotransposon Reverse Transcriptases

Jennifer T. Miller and Stuart F.J. Le Grice*

1. Introduction

With the exception that their life cycle does not involve an extracellular component, events whereby the single-stranded RNA genome
of long terminal repeat (LTR)-containing retrotransposons of
Saccharomyces cerevisiae is converted into integration-competent
double-stranded DNA can be considered analogous to that of animal
retroviruses, whose key enzyme, reverse transcriptase (RT), has been
the subject of intense investigation for almost 50 years. However, the
availability of purified recombinant LTR-retrotransposon RT, and
model nucleic acid substrates mimicking critical steps in reverse transcription, has illuminated several nuances of the LTR-retrotransposon
enzyme. With the availability of high-resolution structural data, the
most prominent of these nuances might be described as a "division of
labor" relative to subunit-associated activity of the human immunodeficiency virus Type 1 (HIV-1) enzyme. The goal of this chapter is
to provide an overview of events supported by the DNA polymerase
and ribonuclease H (RNase H) activities of the Ty3 enzyme that yield

*Reverse Transcriptase Biochemistry Section, Basic Research Laboratory, National
Cancer Institute, Frederick, MD 21702, USA.

double-stranded preintegrative DNA, comparing these with counter-part LTR-retrotransposon and retroviral enzymes.

2. Background/Historical Perspective

Reports from 1970 from the Baltimore and Temin groups of an enzyme from the Rauscher murine leukemia virus (R-MLV[1]) and Rous sarcoma virus (RSV[2]), respectively, with RNA-dependent DNA polymerase, or reverse transcription, activity might be regarded as laying the foundations of retrovirology. Although this could not be predicted at the time, retroviral RT eventually emerged as a tool central to biotechnological advances in the form of cDNA synthesis. Retroviral RT was further propelled into the headlines as an antiviral drug target following identification of the human immunodeficiency virus (HIV) as the etiological agent of acquired immunodeficiency syndrome (AIDS) in 1983.[3,4] Successes in antiretroviral therapy have reflected more than three decades of intense biochemical, biophysical, and structural studies on HIV-1 RT, and along the way have provided valuable insights into aspects of its avian and murine counterparts.

As our understanding of retroviral RT advanced, the Sandmeyer group reported similar organizational features of Ty3, an LTR-retrotransposon of *Saccharomyces cerevisiae*.[5] Common features included (i) LTR sequences terminating in conserved inverted repeats, (ii) a primer binding site (PBS) complementary to a host tRNA from which minus strand DNA synthesis would initiate, and (iii) a polypurine tract (PPT) as a potential primer for second or plus strand DNA synthesis (Fig. 1). The Ty3 polymerase (*pol*) open reading frame also shared the protease/reverse transcriptase-ribonuclease H/integrase organization of many retroviruses.[5] Taken together, these observations might suggest that, mechanistically, RT-mediated events in this LTR-retrotransposon would simply parallel features of more extensively studied retroviruses. In contrast, we and others uncovered unique features of *cis*-acting sequences on the (+) RNA governing their recognition by Ty3 RT, as well as an unexpected organization of the enzyme itself, where DNA polymerase and

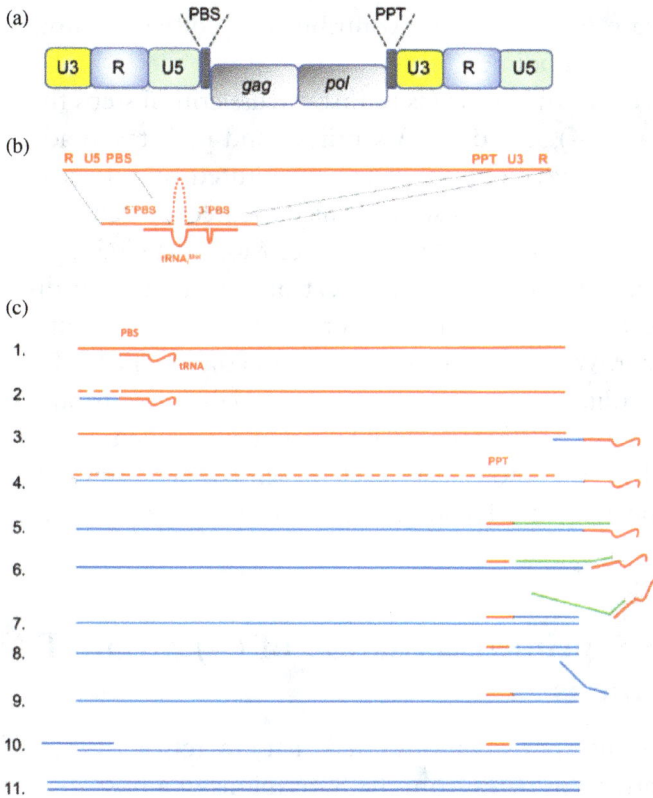

Figure 1. Ty3 preintegrative DNA synthesis. (A) Domain structure of double-stranded preintegrative Ty3 DNA. U3, unique 3′ sequence; R, repeat sequence; U5, unique 5′ sequence; PBS, primer binding site; PPT, polypurine tract. (B) Genomic RNA. The PBS comprises sequences from both the 5′ PBS and 3′ U3 regions. (C) Reverse Transcription Process. (1) Simplified initiation complex, (2) (−) strand strong-stop DNA synthesis, with concomitant RNase H-mediated degradation of (+) RNA. Nascent DNA is shown in blue. RNase H cleavage is shown as broken lines. (3) (−) strand transfer. (4) continued (−) strand synthesis and concomitant degradation of (+) RNA. (5) (+) strand, PPT-primed DNA synthesis (green) extends into tRNA. (6) RNase H-mediated removal of the PPT from (+) strand DNA and tRNA from (−) strand. (7) Second (+) strand DNA (blue) displaces first. (8) PPT removal from (+) DNA. (9) Third (+) strand synthesis initiates and displaces second (+) strand. (10) Second (+) strand transfers to 5′-end of nascent molecule (3′-end of (−) DNA) and PPT is cleaved. (11) Synthesis of both (+) and (−) strands is completed, yielding double-stranded preintegrative DNA. Note: steps 7–9 are not observed in retroviruses.[65]

ribonuclease H activities are contributed by different subunits of an asymmetric homodimer.

The goal of this review is to first discuss critical steps that mediate Ty3 (−) and (+) strand DNA synthesis and have taken advantage of both mutant enzymes and analog-substituted nucleic acid duplexes which, together, will illustrate a high degree of "communication" between the enzyme and its substrate. More detailed aspects of the catalytic centers of Ty3 RT will next be presented, and discussed in the context of the 3-dimensional structure of the asymmetric Ty3 RT homodimer. While the focus will be primarily on Ty3 RT, for which a high-resolution structure is now available, contributions from related LTR-retrotransposon RTs will be included. Although not examined at the molecular level presented here, several excellent reviews on related L1 and LINE elements[6–10] are recommended to readers.

3. tRNA-primed Initiation of (−) Strand DNA Synthesis

Several studies have demonstrated that, in retroviruses, (−) strand DNA synthesis initiates from the 3′ terminus of a host-derived tRNA that shares ~18 nucleotides of complementarity with the PBS of the RNA genome immediately adjacent to the 5′ LTR. Examples of tRNA primer usage include tRNALys, 3 in HIV, tRNATrp in RSV, and tRNAPro in MLV.[11] LTR-retrotransposons have provided interesting variations to this event, including tRNA-independent initiation of (−) strand DNA synthesis in *Schizosaccharomyces pombe* Tf1,[12] and initiation from within the anticodon domain of tRNAiMet in Copia retrovirus-like particles in *Drosophila* and *S. cerevisiae* Ty5, necessitating RNaseP cleavage of the tRNA.[13,14]

Ty3 has provided another variation of "flexibility" of tRNA primer use in LTR-retrotransposons, as a genome sequence adjacent to the 5′ LTR revealed only 8 nucleotides of complimentary to the 3′ terminus of its replication primer, tRNA$_{iMet}$.[15] *In vitro* studies indicated that tRNA$_{iMet}$ failed to stably hybridize with this truncated PBS, from which Gabus *et al.*[15] speculated that additional PBS nucleotides

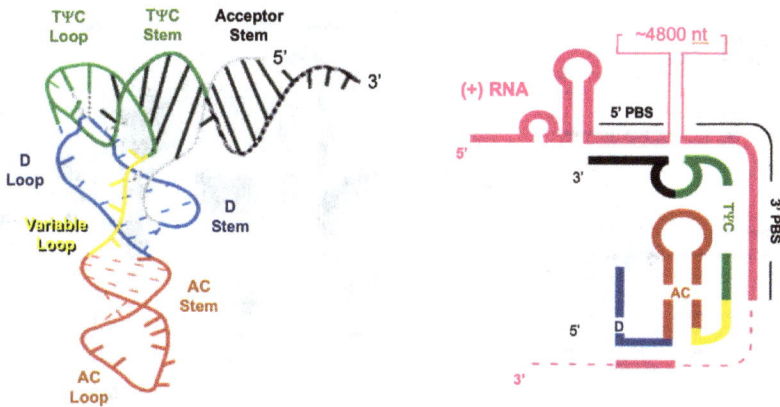

Figure 2. tRNA$_{iMet}$-primed (–) strand DNA synthesis from the Ty3 bipartite PBS. *Left*, cartoon of a generalized tRNA structure wherein the individual stem-loops have been color coded. D, dihydrouridine; AC, anticodon; Ψ, pseudouridine. *Right*, proposed model for initiation of Ty3 (–) strand DNA synthesis from a bipartite PBS. The (+) strand RNA genome is depicted in magenta, while an equivalent color coding as on the left has been employed for tRNA$_{iMet}$.

might be contributed by a distal region of the Ty3 genome. Indeed, subsequent computer alignment analysis uncovered regions of complementarity to the tRNA TψC and D-loops almost 4800 nt downstream, giving rise to the model of Fig. 2.

While the entire Ty1 PBS is located immediately adjacent to the 5′ LTR, it too comprises two segments. In this case, they are separated by a short internal loop.[16] In addition, a long-range pseudoknot interaction between nts 1–7 and 264–270/262–256 and 318–324 has been shown to be necessary for efficient full-length cDNA production.[16,17]

4. RNase H-mediated (–) Strand DNA Transfer

tRNA$_{iMet}$-primed (–) strand DNA synthesis proceeds to the 5′ terminus of the Ty3 (+) RNA genome, concomitant with which RNA of the ensuing RNA/DNA hybrid is degraded by RT-associated RNase H activity. Continued (–) strand DNA synthesis requires transfer of nascent DNA to the 3′ end of the genome, mediated by R region

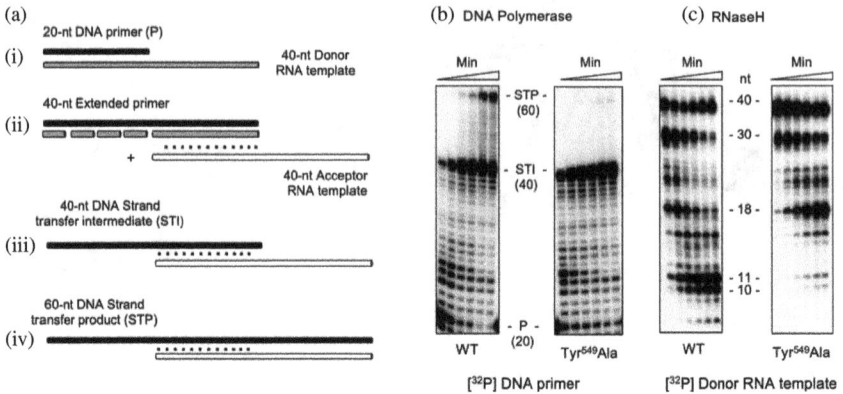

Figure 3.   Ty3 (−) strand DNA transfer and the requirement for RNase H activity. (a) Model substrate (i). RNA-dependent DNA synthesis extends the 20-nt DNA primer to the 5′ terminus of the 40-nt donor RNA template, concomitant with RNA of the nascent RNA/DNA hybrid being degraded (ii). In the presence of an acceptor RNA template sharing 20 nt of complementarity to the donor (iii), transfer of nascent, 40-nt DNA acceptor template and continued RNA-dependent DNA synthesis yields the 60-nt strand transfer product (iv). (b) Monitoring DNA synthesis by wild-type Ty3 RT (WT) and the RNase H mutant Tyr[549]Ala via a 5′-[^{32}P]-labeled DNA primer P, primer, STI, strand transfer intermediate, STP, strand transfer product. (c) RNase H activity of WT and mutant RTs is monitored with a 5′-[^{32}P]-labeled donor RNA template. Incubation time ranged from 0 to 60 min.

homology (see Fig. 1), via a "strand transfer" event that likely involves the Ty3 nucleocapsid protein.[18] While the mechanistic basis remains to be elucidated, the model system shown in Fig. 3 allows monitoring of strand transfer with respect to donor primer extension (Fig. 3(b)) and template degradation (Fig. 3(c)).

Although wild-type Ty3 RT efficiently extends the DNA primer, accumulation of the strand transfer product is delayed, correlating with the appearance of predominant donor template hydrolysis fragments of 11 and 10 nt. This combination polymerization and degradation event would, as a consequence, make ~10 nt of the donor template available for hybridization of the acceptor, after which displacement/dissociation of residual template fragments would allow complete hybridization of the latter. The extent to which the donor template must be degraded becomes apparent when analyzing the effect of point mutations in the RNase H domain. Figures 3(b) and

3(c) indicate that RNA-dependent DNA polymerase activity of Ty3 RT mutant Tyr[549]Ala is equivalent to the wild-type enzyme, yet it supports negligible strand transfer activity (Ty3 RT residue Tyr[549] is equivalent to His[539] of HIV-1 RT, which is important for RNaseH catalysis[19,20]).

Unimpaired primer extension predicts both enzymes are positioned at the 5′ terminus of the donor template to support strand transfer (these seem like two separate points). However, an intriguing observation is that RNase H hydrolysis products from the mutant enzyme are quantitatively different, revealing a principal minimum size length of 18 nt. The length of this donor template fragment would span almost the entire homology region with the acceptor template (20 nt in the model of Fig. 3(a)), thereby impairing annealing of the latter and, as a consequence, an inability to support DNA strand transfer. Our previous work with HIV-1 RT identified the mutant p66/p51D13, which likewise retained high levels of polymerization-independent RNase H activity yet failed to support significant strand transfer activity.[21] While DNA strand transfer is the outcome of coordinated polymerization and RNase H activities, studies here with Ty3 RT highlight an intricate series of events underlying this "specialized" RNase H-mediated event.

5. Polypurine Tract (PPT)-primed (+) Strand DNA Synthesis

Unlike retroviruses, whose (+) strand PPT primer comprises homopolymeric runs of A and G residues (e.g., $5′ - (A)_4\text{-}U\text{-}(A)_4\text{-}(G)_6 - 3′$ for HIV-1), the Ty3 PPT comprises mostly alternating adenine and guanine residues (Fig. 4). Despite this, model systems have successfully recapitulated its excision from genomic (+) RNA and nascent (+) DNA,[19] suggesting that unique structural features of the RNA/DNA hybrid might play a role in correct positioning of the retrotransposon enzyme. Earlier structural work of Kopka *et al.*[22] raised the possibility that altered stacking at the -A-G-A- step within the HIV-1 PPT introduced a local deformation that could contribute to its recognition by the retroviral enzyme. Several lines of evidence suggest that localized

(a)   (-) DNA   (b)

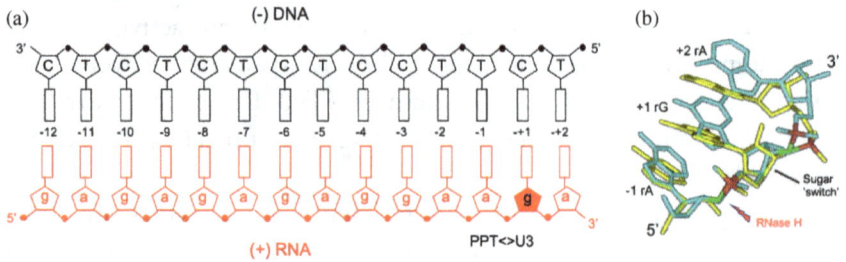

(+) RNA   PPT<>U3

Figure 4.   (a) Sequence of the Ty3 PPT RNA/DNA hybrid. DNA and RNA nucleotides are denoted in black and red, respectively. Base pair numbering is relative to the PPT/U3 cleavage junction. RNA residue +1 g, where structural alterations have been observed by both NMR spectroscopy and sensitivity to chemical acylation, is highlighted.[24] (b) A sugar pucker switch affects RNA backbone geometry at the Ty3 PPT/U3 junction. The RNA strand -rA-rG-rA- step at the junction is shown with standard A-form geometry (all ribose sugars have a C3-endo conformation) as a blue stick representation. The yellow stick representation of this step is modeled with the ribose of residue +1rG in a C2'-endo sugar pucker conformation and riboses of –1rA and +2rA in C3'-endo conformations. In both models, backbone 3'-oxygens and phosphates are shown in green and red, respectively. Scissile bond and +1rG sugar switch are indicated. Adapted with permission from Ref. [23].

deformations within the Ty3 PPT provide some form of long-range structural coupling. As an example, we reported an A- to B- transition in sugar pucker of +1 g at the Ty3 PPT/U3 junction, which would be predicted to alter the backbone conformation of the RNA/DNA hybrid downstream of this residue.[23] In contrast, the conformation of the phosphodiester backbone 5' to this residue is not perturbed by the sugar pucker switch (Fig. 4(b)). By combining selective 2'-hydroxyl acylation with mass spectrometry (SHAMS[24]), acylation of +1g ribose could be observed, again suggestive of local conformational fluidity. Secondly, a combination of NMR spectroscopy, mass spectrometry, and isothermal titration calorimetry has identified the PPT/U3 junction as a preferential binding site for the aminoglycoside neomycin B.[25] Thirdly, single-site T to 2,4-difluorotoluene (F) substitutions of the Ty3 PPT affected sugar puckering for nucleotides well removed from the site of analog insertion. As an example, substituting DNA base –11 T with F produced a periodic switch in sugar pucker for upstream RNA nucleotides –9, –7, –5, –3, and –1, suggesting long-range structural coupling.[24] These combined observations suggest a model where RNase H cleavage fidelity may be mediated by local distortion of the RNA/DNA hybrid in the immediate vicinity of the PPT/U3 junction.[23]

A novel approach to better understand communication between the LTR retrotransposon RT and its nucleic acid substrate is through the strategy of nucleoside analog interference, an example of which is presented in Fig. 5. 2,4-difluorotoluene (Fig. 5(a)) is a nonpolar size and shape mimic of thymine, but with severely reduced hydrogen bond formation with adenine.[26,27] F thus provides a good test for the importance of hydrogen bonding groups on the fidelity of PPT selection. When positioned within (–) DNA of the HIV-1 PPT, dual T -> F substitutions relocated the position of RNase H cleavage 3–4 bp ahead of their site of insertion, suggesting the locally destabilized region was "sensed" or "sequestered" by a structural component of the retroviral enzyme.[28] When the equivalent strategy was applied to Ty3 RT,[29] Fig. 5(b) indicates that replacing template nucleotide –2T with F promoted cleavage ~12 bp downstream of the site of insertion (+10), in addition to the PPT/U3 junction. This effect was more pronounced when template nucleotides –1T/–2T were replaced with F, where the PPT/U3 junction was virtually bypassed in favor of cleavage at positions +10/+11. In the absence of crystallographic

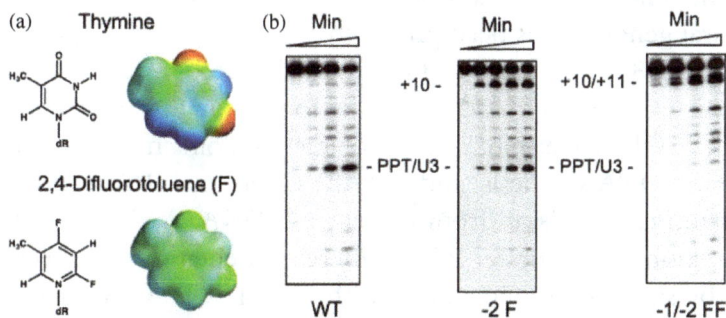

Figure 5. Deciphering Ty3 PPT cleavage specificity via nucleoside analog mutagenesis. (a) Chemical structure of the DNA base thymine and its non-hydrogen bonding isostere, 2,4-difluorotoluene (F). The chemical structure of each is presented alongside a space-filled molecule. For the latter, electrostatic surface potentials are highlighted in red (negative) and blue (positive). (b) Replacing PPT DNA template nucleotides –2T or –1T/–2T with F redirects RNase H cleavage from the PPT/U3 junction to a novel site ~12 bp downstream.

data, the results of Fig. 5(b) made important predictions concerning the disposition of Ty3 RT subdomains on the RNA/DNA hybrid. By analogy with HIV-1 RT,[30] ~12-bp separation between the sites of analog insertion and hybrid cleavage reflects the spatial separation of the Ty3 thumb subdomain and RNase H active site, a notion that would be borne out later by X-ray crystallography.[31] Localized disruption of hydrogen bonding thus appears to provide a region with increased malleability that promotes an interaction with the thumb subdomain.

Can any biological function be extrapolated from such nucleoside analog interference experiments? In addition to recognizing the PPT/U3 junction, NMR studies of Brinson *et al.*[25] revealed a second neomycin B binding site immediately upstream of the Ty3 PPT, an observation borne out by ESI-FTICR mass spectrometry.[24] While further experimentation is necessary, such studies point toward structural anomalies located at both ends of the Ty3 PPT driving its recognition and faithful processing by domains of the retrotransposon RT. Curiously, while accurate Ty1 PPT processing has been recapitulated by Wilhelm *et al.*,[32,33] it deviates significantly from the "all-purine" notion in that it comprises short runs of purines interspersed by a pyrimidine. In the model proposed by these authors, a U-rich segment immediately upstream of the PPT and common to many retroviral genomes has been proposed as a recognition element.

The Ty3 PPT/U3 junction must be recognized by (i) the C-terminal RNase H domain, in order to provide the (+) strand RNA primer, (ii) the N-terminal DNA polymerase domain for initiation of (+) strand DNA synthesis, and (iii) the C-terminal RNase H domain for its ultimate excision from nascent (+) DNA. A clue to how the same region can sequester RT in different orientations has been provided by single-molecule studies with HIV-1 RT.[34-37] Among these, Abbondanzieri *et al.*[34] demonstrated substrate-dependent alterations in enzyme binding orientation. An oligonucleotide duplex simulating a PPT which had not been cleaved at its 3′ terminus bound RT in an orientation positioning the RNase H domain over the PPT/U3 junction. In contrast, an RNA/DNA hybrid simulating early steps of DNA-dependent DNA synthesis induced enzyme binding in the opposite orientation. Enzyme orientation could also be modulated by incubation with either DNA polymerase[34] or RNase H inhibitors.[35] Collectively, such observations suggest a considerable degree of communication between HIV-1 RT and its substrate which undoubtedly can be extended to its LTR-retrotransposon counterpart.

Although we have concentrated here on subtle structural features of the PPT that mediate its recognition by Ty3 RT, the reader is referred to several articles from the Wilhelm group addressing PPT selection in the LTR-retrotransposon Ty1.[32,33,38,39]

6. Role of the Ty3 RT Thumb Subdomain in Substrate Recognition

While the previous section might lead one to propose a role for the Ty3 RT thumb subdomain as a "sensor" of nucleic acid geometry, early delineation of important structural motifs has benefitted from a combination of nucleoside analog and/or site-directed mutagenesis. As an example of the former, and based on crystallographic analysis of HIV-1 RT indicating contacts between a-helix H and the nucleic acid duplex 3–6 bp downstream of the DNA polymerase catalytic center,[30] locked nucleic acids (LNAs, Fig. 6(a)) can provide insights into the consequence of reducing duplex flexibility on critical contacts with the LTR-retrotransposon enzyme.[40] Figure 6(b) demonstrates that single nucleotide extension activity is severely reduced on DNA duplexes containing LNA substitution of primer nucleotides –3 and –4, partially restored when primer nucleotide –5 is replaced, and unaffected by –6 and –7 insertions. Conversely, LNA replacement of template nucleotides –3, –4, and –5 has minimal impact on DNA polymerase activity, and their substitution at positions –6 and –7 is inhibitory. Taken together, data of Fig. 6(b) predict that the geometry of the nucleic acid duplex 3–7 bp downstream of the primer terminus is critical for interaction with the thumb subdomain. Consistent with this notion is our observation that primer extension activity is severely impaired when a tetrahydrofuran linkage (which removes the nucleobase but preserves the sugar-phosphate backbone) is introduced at primer positions –3 and –4.[40] These and related

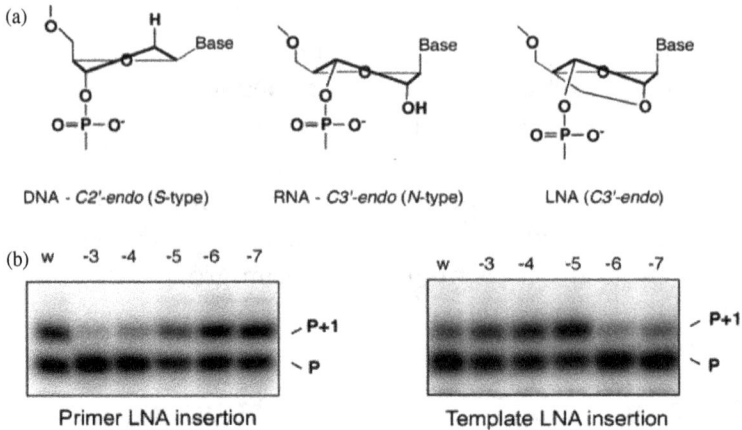

Figure 6.   (a) *C3'-endo* sugar pucker of locked nucleic acid (LNA) compared with that of DNA *(C2'-endo)* and RNA (*C3'-endo*). The C2' – C4' bridge "locks" the LNA ribose in the 3'-endo conformation common to A-form duplexes. (b) Effect of LNA substitutions of primer and template nucleotides on single nucleotide extension activity of Ty3 RT. The position of LNA insertion is relative to the primer 3' OH of the DNA duplex. w, wild-type substrate. Notations –3 to –7 refer to the site of LNA insertion in the template or primer.

experiments[29,41,42] illustrate how high-resolution structural data can be obtained, even in the absence of crystallographic information.

With respect to critical residues of the Ty3 RT thumb, the structure of HIV-1 RT complexed with duplex DNA[30] has impicated *a*-helix H residues Gln[258], Gly[262], and Trp[266] in contacting the sugar phosphate backbone of the primer 3–6 bp downstream of the DNA polymerase catalytic center, and Asn[265] with contacting the backbone of template nucleotides –6/–7. Sequence comparison of LTR-retrotransposons and retroviruses combined with modeling studies[40] suggested Gln[290], Gly[294], Tyr[298] and Asn[297] as their Ty3 RT counterparts. Under conditions where DNA synthesis was limited to a single enzyme binding event, alanine scanning mutagenesis showed near wild-type activity for mutants Gln[258]Ala and Asn[297]Ala, but significantly reduced activity when Gly[294] and Tyr[298] were substituted.[40] Surprisingly, a Gly[294]Ala mutation also inhibited sequence-independent RNase H activity as well as specific cleavage at the PPT/U3 junction. When the same mutant was challenged with nucleic acid duplexes containing tetrahydrofuran linkages, activity comparable with that on the unsubstituted duplex was restored with duplexes whose primer was modified at positions −4 and −5. Modeling studies with unsubstituted DNA suggested that the methyl group of the Ala side chain was oriented toward the stacking interface between nucleobases −4 and −5. Eliminating the primer nucleobase via tetrahydrofuran was proposed as a means of suppressing steric interference introduced by a Gly^{294}Ala substitution, arguing that the Ty3 RT Gly^{294} main chain mediates contacts with primer nucleobase −4. Although we again caution that nucleoside analog mutagenesis data give only clues to sites of protein–nucleic acid contact, the solution structure of Ty3 RT in the presence of an RNA/DNA hybrid (see later) has confirmed the importance of thumb residues Phe^{292} and Gly^{294}, the latter of which has been suggested to make a critical contribution to this component of the subunit interface.[43]

7. DNA Polymerase and RNase H Active Sites

7.1 DNA polymerase

Sequence homology, combined with structural comparisons between nucleic acid polymerases, defined Asp^{151}, Asp^{213}, and Asp^{214} (the counterparts of HIV-1 RT p66 residues Asp^{110}, Asp^{185}, and Asp^{186}, respectively) as DNA polymerase active site residues coordinating the two divalent Mg^{++} ions essential for catalysis[44] (see Fig. 7). Site-directed mutagenesis suggested Asp^{151} as the most critical residue of the triad, since its substitution with either Asn or Glu had major implications for DNA polymerase, RNase H, and pyrophosphorolysis activity.[44] The notion that such pleiotropic effects reflected diminished substrate binding was ruled out, as wild-type RT and Asp^{151}Glu displayed nanomolar affinities for duplex DNA.[44] Since Ty3 RT mutant Asp^{151}Glu could support primer extension, but only by a single nucleotide, it led us to propose a translocation defect reflected by an inability to release pyrophosphate following phosphodiester bond formation. Although this needs to be validated experimentally, an outcome of

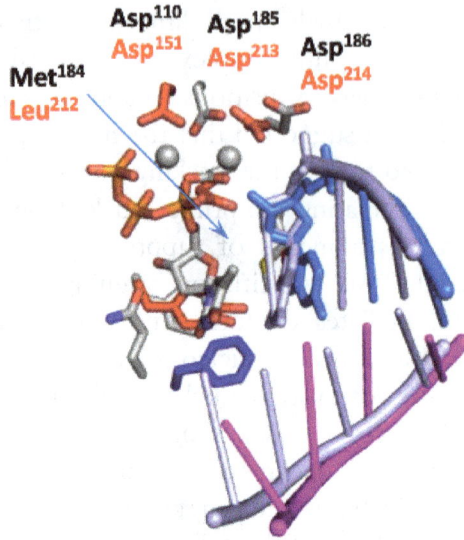

Figure 7.   DNA polymerase active site residues of Ty3 (red) and HIV-1 RT (black). Catalytic Mg++ ions and the incoming dTTP are in gray and dark gray, respectively. HIV-1 duplex DNA strands are shown as a light blue ladder, and the RNA template and DNA primer bound by Ty3 RT are in magenta and marine, respectively. Adapted with permission from Ref. [37].

this defect would be to trap the primer terminus at the pre-translocation (N) site defined by Sarafianos *et al.* for HIV-1 RT.[45]

Although loss of activity when Asp[213] was replaced with Asn was predicted, activity of the Asp[213]Glu mutant was several orders of magnitude higher than its HIV-1 RT counterpart.[46] Despite retaining DNA polymerase activity, Asp[213]Glu RT failed to catalyze pyrophosphorolysis. The combined phenotypes of Asp[151] and Asp[213] mutants support their involvement in Mg++ coordination. Unaltered DNA polymerase and pyrophosophorolysis activity when Asp[214] was substituted with Asn or Glu indicates that the second aspartate of the Ty3 RT -Y-X-D-D- motif is less essential. Alternative roles for Asp[214] in active site architecture (based on studies with F29 DNA polymerase[47]), or supporting the negative charge within the catalytic site to facilitate translocation (suggested by Sarafianos *et al.*[45]), have been proposed. However, when Asp[214] mutations were introduced into the RT domain of pEGTy3-1, a plasmid harboring a replication-competent Ty3 element, all Asp[214] substitutions tested were inhibitory for transposition.

These strains could be rescued by co-transfection of a plasmid containing the parental RT gene, thus correlating loss of transposition activity directly to alterations in the Ty3 RT DNA polymerase active site.[44] The power of the Ty3 genetic approach thus indicates that slight reductions in activity evidenced through biochemical evaluation of the recombinant enzyme likely have a cumulative effect on the multiple and sophisticated steps of Ty3 reverse transcription. Ty1 RT activity is likewise Mg^{++}-dependent and, facilitated by invariant aspartate active site residues, Asp^{129}, Asp^{210}, and Asp^{211}.[48] While mutations at the first two active site residues are inactive, Ty1 RT mutant $Asp^{211}Asn$ exhibits altered pyrophosphate binding and release, as well as a preference for Mn^{++} over Mg^{++}.[49] Intriguingly, suppressor mutations obtained that increase WT RT activity in Mn^{++} occur in the RNase H domain, but do not alter RNase H activity, indicating subtle cross talk between the two domains.[50]

7.2 RNase H

Structurally, retroviral DNA polymerase and RNase H domains are separated by an ~100 residue "connection" subdomain (Fig. 8(b)).[30] A major difference between the LTR-retrotransposon and retroviral enzymes is lack of this domain in Ty3 RT, thereby positioning the catalytic centers considerably closer. Sequence homology between cellular, retroviral, and LTR-retrotransposon RNases H[51] defined a consensus -Asp-Glu-Asp-Asp- active site motif, which for Ty3 RT comprises Asp^{358}, Glu401, Asp^{426}, and Asp^{469}.[19] Model RNA/DNA hybrid substrates have indicated that, in the absence of DNA synthesis, wild-type Ty3 RT cleaves the substrate ~13 and 21 bp downstream of the primer terminus.[51,52] Substituting any residue of the -Asp-Glu-Asp-Asp- motif severely impairs RNase H activity, while at the same time having minimal effect on DNA polymerase function.[19] As outlined earlier, selective loss of RNase H function leads to a breakdown in DNA strand transfer activity. As would be predicted from biochemical investigation, none of these mutants was competent for transposition.

Tisdale *et al.* highlighted His^{539} of the "flexible His-loop" as essential for RNase H activity of HIV-1 RT *in vitro*, and virus

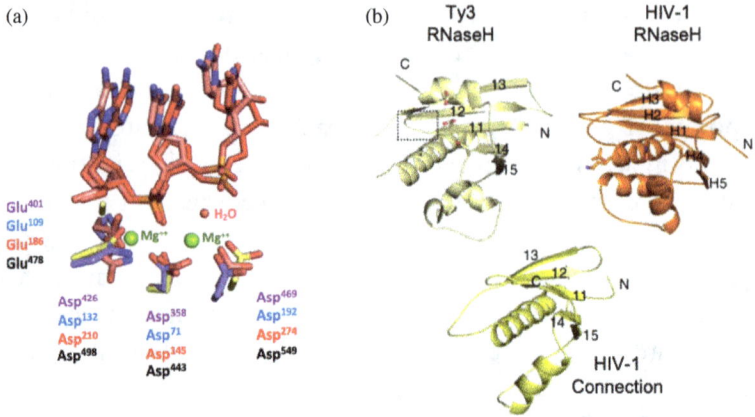

Figure 8. (a) RNase H active site residues of Ty3 RT (magenta), *Bacillus halodurans* RNase H1 (blue), and human RNase H1 (red). RNA strands from human and bacterial RNases H1 are shown in salmon and red. Although carboxylates of HIV RNase H are not depicted in the cartoon, counterpart residues of the -D-E-D-D-motif are indicated in black. (b) Comparison of RNase H and connection domains. Cartoon representations of Ty3 RNase H (*Upper left*), HIV-1 RNase H (*Upper, right*), and the HIV-1 RT p66 connection subdomain, an "RNase H remnant" (*Lower*). Strands of the central α-sheet of each domain are indicated. Adapted with permission from Ref. [37].

replication *in vivo*.[20] Aligning catalytic residues in the RNase H domains of the *gypsy* group of retrotransposons and plant caulimoviruses[19] indicated conservation of a Tyr residue at the equivalent position. In the case of Ty3 RT, it is Tyr[459]. Despite retaining significant levels of DNA polymerase and RNase H activity (see Fig. 3), a recombinant Ty3 element harboring a Tyr[459]Ala mutation failed to support transposition.[19] Thus, while the general process of degrading the RNA/DNA reverse transcription intermediate remains, such data indicate that there are steps in the replication cycle (e.g., DNA strand transfer and PPT selection/removal) that require more precise processing by the RNaseH domain.

8. Bringing Things Together: Crystal Structure of a Ty3 RT-RNA/DNA Hybrid

Approximately 14 years after providing the first detailed biochemical analysis of recombinant Ty3 RT,[51] a collaboration between our group and that of Marcin Nowotny, International Institute of Molecular and Cell Biology, Warsaw, Poland, produced a structure of Ty3 RT complexed with its PPT-containing RNA/DNA hybrid at 3.5Å resolution.[43] At the same time, we provided high-resolution structures of HIV-1 and xenotropic murine leukemia virus-related virus (XMRV) RTs complexed with an RNA/DNA hybrid,[52,53] allowing us to compare and contrast the LTR-retrotransposon with its monomeric XMRV and dimeric counterparts. Rather than fall into one of these two structural classes, Ty3 RT has proven to be quite unique, both structurally and functionally.

Early size exclusion chromatography studies in the absence of nucleic acid substrate[51] suggested a 55-kDa monomer as the active form of Ty3 RT, which at the time was in keeping with structural data on the XMRV enzyme. In contrast, crystallographic data of the Ty3 enzyme:substrate complex predicted a homodimeric organization. Surprisingly, analytical ultracentrifugation, sedimentation velocity analysis suggested *substrate-induced* dimerization, contrasting with HIV-1 RT which adopts both a homo- and heterodimeric configuration in the absence and presence of substrate.[30,53] Based on these studies, the subunits of Ty3 RT have been designated A and B, the subdomains of which are structurally similar. A cartoon depiction of Ty3 RT in the presence of the short PPT RNA/DNA hybrid is provided in Fig. 9(a), and contacts to the substrate in Fig. 9(b). The subunit B RNase H domain is positioned between its fingers and palm, blocking the substrate binding cleft and causing the thumb subdomain to be displaced from the palm and rotated relative to the RNase H domain. Thus, and as shown in Fig. 9(a), subunit A provides the DNA polymerase catalytic center.

The availability of a crystal structure for the Ty3 RT complexed with an RNA/DNA hybrid allowed us to compare and contrast it with previous proposals from biochemical studies. For example, DNaseI/nuclease S1 footprinting analysis with duplex DNA suggested an interaction with template nucleotides +7 to −24 and primer nucleotides −1 to −25.[51] Modeling a longer duplex onto the Ty3 crystal structure raised the possibility of an interaction with the positively charged region of the subunit B thumb, supporting nuclease footprinting data and suggesting that the thumb could promote

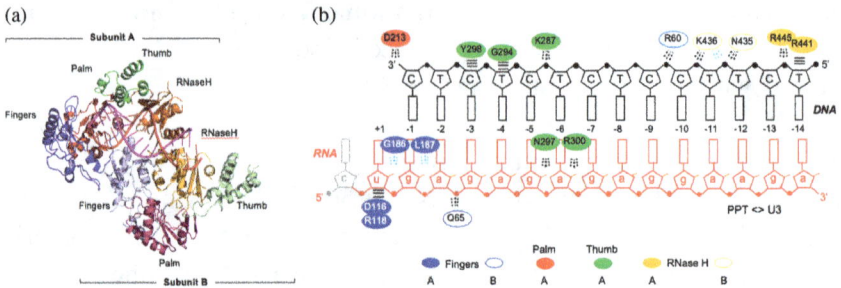

Figure 9. (a) Structure of the Ty3 p55 RT homodimer complexed with its PPT RNA/DNA hybrid. Individual subunits have been designated A and B, and subdomains are color coded according to those of HIV-1 RT, i.e., blue (fingers), red (palm), green (thumb), and gold (RNase H). In contrast to the HIV-1 enzyme, Ty3 RT lacks a copy of the connection subdomain. Subunit A subdomains are represented by darker shading, and those of subunit B by lighter shading. DNA and RNA strands of the PPT are depicted in magenta and red, respectively. Note that, despite each subunit retaining an RNase H domain, the asymmetric organization of the Ty3 RT homodimer closely resembles that of the p66/p51 HIV-1 RT heterodimer whose p51 subunit lacks this domain. (b) Summary of protein–nucleic acid interactions. The PPT/U3 junction is indicated, and the 5′ RNA nucleotide absent in the structure is shown in gray. Ovals are colored by protein domain as in (a), with solid and empty ovals denoting subunit A and B, respectively. Parallel horizontal lines indicate van der Waals interactions. Diagonal and vertical lines indicate interactions mediated by the backbone (cyan) or side chains (black) of the protein. Adapted with permission from Ref. [37].

further stabilization of the RNA-DNA substrate. Curiously, template and primer nucleotides between positions −16 and −18 were rendered hypersensitive to DNAse I cleavage. This "window of accessibility" is akin to hydroxyl radical footprinting observations of Metzger *et al.* with HIV-1 RT,[54] differing inasmuch as the latter defined a window between nucleotides −8 and −11, Although more experimentation is necessary, these studies collectively raise the notion of local enzyme-induced alterations in substrate conformation, either between the DNA polymerase and RNase H domains (HIV-1 RT) or immediately adjacent to the RNase H domain (Ty3 RT). Further, interactions mediated by the subunit A thumb subdomain supported biochemical studies showing the importance of Phe[292], Gly[294], and Tyr[298].[40] Gly[294]Ala RT was the most affected in the absence of a heparin trap, indicating its critical contribution to this component of the dimer interface. Experiments with LNA-substituted nucleic acids (see Section 5) also predicted interactions of subunit A thumb residues Tyr[298] and Gly[294] with DNA nucleotides –3 and –4, supporting and extending findings from mutagenesis analysis of HIV-1 RT.[55,56] Thus, while precise details of their contribution cannot be elucidated, such experiments illustrate the power of nucleoside analog mutagenesis in highlighting critical protein–nucleic acid interactions.

While DNA polymerase activity could readily be ascribed to RT subunit A from the crystal structure, applying the same approach to the functional RNase H domain was not possible, since neither RNase H active site interacted with the RNA strand of the RNA/DNA hybrid in the crystal structure. This predicted that a significant conformational change would be necessary to position either active site over the scissile bond. Modeling studies suggested proximity of the subunit B RNase H domain to the scissile phosphate, and its movement could be accommodated by a translation of ~40 Å without invoking severe clashes, while preserving dimerization contacts of the palm and fingers subdomains. When the same strategy was applied to subunit A, aligning its active site over the scissile bond was predicted to disrupt the dimer structure and eliminate critical contacts between its thumb subdomain and the substrate. The implication of this modeling exercise suggested a unique "division of labor" between the two Ty3 RT subunits, i.e., DNA polymerase activity resides with subunit A and RNase H activity with subunit B.[43]

Answering this challenge biochemically exploited phenotypic mixing experiments, based on structural data on residues of the RNase H active site and the dimer interface. Arg[140] and Arg[203] mediate critical contacts at the dimer interface of subunit A, but are distal from the dimer or substrate interface in subunit B. Thus, when an Arg[140]Ala/Arg[203]Ala, dimerization-deficient mutant[43] was mixed with an RNase H-deficient Asp[426]Asn mutant in the presence of substrate, the only *dimer* combinations would be an Asp[426]Asn mutant lacking RNase H activity and a mixed dimer whose subunit B contributed Arg[140]Ala/Arg[203]Ala. If RNase H activity was a function of subunit B, only the latter mixed homodimer would display activity. When these two mutant enzymes were reconstituted, RNase H activity could indeed be rescued, confirming it as a function of subunit B.[43]

9. Topological Similarities between Ty3 and HIV-1 RT

In addition to the "division of labor" with respect to enzymatic activities, a major difference between the LTR-retrotransposon and retroviral RTs is the absence of a connection subdomain located between the DNA polymerase and RNase H domains of the Ty3 enzyme, i.e., its RNase H domain is positioned immediately adjacent to the thumb. Through their phylogenetic analysis of RNase H domains, this has spawned the proposal of Malik and Eickbush[57] that retroviral RTs may have evolved from LTR retrotransposon enzymes by converting their RNase H domain to a connection with concomitant loss of catalytic function and subsequent recruitment of a new RNase H1 domain from either a eukaryotic host genome or a non-LTR retrotransposon (see Fig. 8). Despite these structural differences, Fig. 10 demonstrates that the "two-RNase H" Ty3 homodimer and "one-RNase H" HIV-1 heterodimer are surprisingly similar topologically. As might be predicted, the fingers-palm-thumb organization of their DNA polymerase domains can be likened to a right hand grasping the nucleic acid duplex. More striking, however, is the overall structural similarity between the HIV-1 RT p51 subunit and Ty3 RT subunit B, where the connection subdomain of the former and RNase H domain of the latter are positioned between the fingers and palm. As a consequence, the quaternary structures of the dimeric enzymes are quite superimposable, and, as proposed by Skalka,[58] the subtle differences likely relate to functions and interactions critical to their replication.

10. A Final Word: Roles for Integrase in Ty3 Reverse Transcription?

Kirchner and Sandmeyer demonstrated that C-terminal deletions of the Ty3 integrase (IN) protein induced a dramatic reduction in the amount of Ty3 DNA *in vivo* and a decrease in RT activity *in vitro*, without affecting the size or amount of the 55-kDa RT in virus-like particles (VLPs).[59] At the same time, these authors identified a

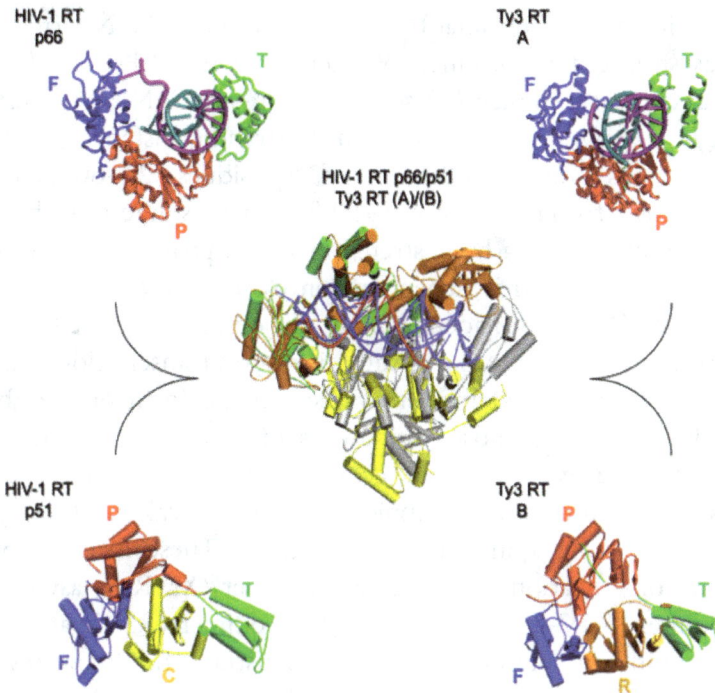

Figure 10. Structural comparison of the p55 Ty3 RT homodimer and p66/p51 HIV-1 RT heterodimer. *Upper,* fingers (F)/palm (P)/thumb (T) of p66 RT and Ty3 RT subunit A in the presence of nucleic acid. Subdomains are color coded as described in Fig. 9 (a). *Lower,* p51 HIV-1 RT compared with Ty3 RT subunit B. Note that the p51 connection subdomain (C) is replaced by the RNase H domain of Ty3 RT (R). *Center,* superimposition of the Ty3 p55 homodimer on the p66/p51 HIV-1 RT heterodimer. HIV-1 RT subunits are depicted in orange and gray, while those of Ty3 RT are in green and yellow.

115-kDa polypeptide in VLPs that reacted with antibodies to IN, suggesting a biological role for an RT-IN fusion protein. Subsequent studies of Nymark-McMahon *et al.* showed that reduced cDNA levels of IN mutants caused reduced levels of early reverse transcription intermediates, such as (–) strand strong-stop DNA.[60,61] This defect could be rescued by expression of a tripartite capsid/RT/IN fusion protein, leading to a model where independent IN domains contribute to both reverse transcription and integration. Likewise, efficient Ty1 reverse transcription requires the presence of IN protein. *In vitro,* a 115 amino acid C-terminal IN fragment fused to the N terminus of RT was sufficient to produce WT activity.[38] In VLPs, IN deletion mutants could synthesize (−) strand strong-stop DNA and perform RNase H cleavage, but were deficient in strand transfer or (+) strand DNA synthesis. Notably, a deletion of IN residues 521–607, a highly acidic region, strongly reduced Ty1 RT activity, suggesting these IN residues interact with a basic stretch in RT to promote correct folding.[62] This interaction is dependent on close association of IN and RT, as when IN was provided *in trans*, the defect in reverse transcription was not rescued.[63] A multi-subunit reverse transcription complex has also been proposed for retroviruses in the form of the ab RT heterodimer of RSV, whose b subunit is a fusion of RT and IN polypeptides.[64] The availability of recombinant RT/IN complexes for reconstitution into such a complex for biochemical and biophysical analyzes would help answer such questions. These issues notwithstanding, organization of RTs as a monomer (XMRV), asymmetric heterodimer (HIV-1), multi-protein heterodimer (RSV), and asymmetric, multi-protein homodimer (Ty3) illustrates that more research is needed to fully understand the multiple roles these highly versatile enzymes must perform during the replication of retroviruses and transposable elements.

Acknowledgments

SFJLG and JTM are funded by the Intramural Research Program of the National Cancer Institute, National Institutes of Health, Department of Health and Human Services.

References

1. Baltimore D. RNA-dependent DNA polymerase in virions of RNA tumour viruses. *Nature* 1970;226:1209–11. doi:10.1038/2261209a0.
2. Temin HM, Mizutani S. RNA-dependent DNA polymerase in virions of Rous sarcoma virus. *Nature* 1970;226:1211–3. doi:10.1038/2261211a0.
3. Barre-Sinoussi F, *et al.* Isolation of a T-lymphotropic retrovirus from a patient at risk for acquired immune deficiency syndrome (AIDS). *Science* 1983;220:868–71. doi:10.1126/science.6189183.

4. Gallo RC, *et al.* Isolation of human T-cell leukemia virus in acquired immune deficiency syndrome (AIDS). *Science* 1983;220:865–7. doi:10.1126/science.6601823.

5. Hansen LJ, Chalker DL, Sandmeyer SB. Ty3, a yeast retrotransposon associated with tRNA genes, has homology to animal retroviruses. *Mol Cell Biol* 1988;8:5245–56. doi:10.1128/mcb.8.12.5245.

6. Banuelos-Sanchez G, *et al.* Synthesis and characterization of specific reverse transcriptase inhibitors for mammalian LINE-1 retrotransposons. *Cell Chem Biol* 2019;26:1095–109 e1014. doi:10.1016/j.chembiol.2019.04.010.

7. Pizarro JG, Cristofari G. Post-Transcriptional control of LINE-1 retrotransposition by cellular host factors in somatic cells. *Front Cell Dev Biol* 2016;4:14. doi:10.3389/fcell.2016.00014.

8. Pradhan M, Govindaraju A, Jagdish A, Christensen SM. The linker region of LINEs modulates DNA cleavage and DNA polymerization. *Anal Biochem* 2020;603:113809. doi:10.1016/j.ab.2020.113809.

9. Taylor MS, *et al.* Dissection of affinity captured LINE-1 macromolecular complexes. *Elife* 2018;7. doi:10.7554/eLife.30094.

10. Viollet S, Monot C, Cristofari G. L1 retrotransposition: the snap-velcro model and its consequences. *Mob Genet Elements* 2014;4:e28907. doi:10.4161/mge.28907.

11. Le Grice SF. "In the beginning": initiation of minus strand DNA synthesis in retroviruses and LTR-containing retrotransposons. *Biochemistry* 2003;42:14349–55. doi:10.1021/bi030201q.

12. Atwood A, Lin JH, Levin HL. The retrotransposon Tf1 assembles virus-like particles that contain excess Gag relative to integrase because of a regulated degradation process. *Mol Cell Biol* 1996;16:338–46. doi:10.1128/mcb.16.1.338.

13. Ke N, Gao X, Keeney JB, Boeke JD, Voytas DF. The yeast retrotransposon Ty5 uses the anticodon stem-loop of the initiator methionine tRNA as a primer for reverse transcription. *RNA* 1999;5:929–38. doi:10.1017/s1355838299990015.

14. Kikuchi Y, Ando Y, Shiba T. Unusual priming mechanism of RNA-directed DNA synthesis in copia retrovirus-like particles of Drosophila. Nature 1986;323:824–6. doi:10.1038/323824a0.

15. Gabus C, *et al.* The yeast Ty3 retrotransposon contains a 5′-3′ bipartite primer-binding site and encodes nucleocapsid protein NCp9 functionally homologous to HIV-1 NCp7. *EMBO J* 1998;17:4873–80. doi:10.1093/emboj/17.16.4873.

16. Bolton EC, Coombes C, Eby Y, Cardell M, Boeke JD. Identification and characterization of critical cis-acting sequences within the yeast Ty1 retrotransposon. *RNA* 2005;11:308–22. doi:10.1261/rna.7860605.

17. Huang Q, *et al.* Retrotransposon Ty1 RNA contains a 5'-terminal long-range pseudoknot required for efficient reverse transcription. *RNA* 2013;19:320–32. doi:10.1261/rna.035535.112.

18. Sandmeyer SB, Clemens KA. Function of a retrotransposon nucleocapsid protein. *RNA Biol* 2010;7:642–54. doi:10.4161/rna.7.6.14117.

19. Lener D, Budihas SR, Le Grice SF. Mutating conserved residues in the ribonuclease H domain of Ty3 reverse transcriptase affects specialized cleavage events. *J Biol Chem* 2002;277:26486–95. doi:10.1074/jbc. M200496200.

20. Tisdale M, Schulze T, Larder BA, Moelling K. Mutations within the RNase H domain of human immunodeficiency virus type 1 reverse transcriptase abolish virus infectivity. *J Gen Virol* 1991;72(Pt 1):59–66. doi:10.1099/0022-1317-72-1-59.

21. Jacques PS, Wohrl BM, Howard KJ, Le Grice SF. Modulation of HIV-1 reverse transcriptase function in "selectively deleted" p66/p51 heterodimers. *J Biol Chem* 1994;269:1388–93.

22. Kopka ML, Lavelle L, Han GW, Ng HL, Dickerson RE. An unusual sugar conformation in the structure of an RNA/DNA decamer of the polypurine tract may affect recognition by RNase H. *J Mol Biol* 2003;334:653–65. doi:10.1016/j.jmb.2003.09.057.

23. Yi-Brunozzi HY, Brabazon DM, Lener D, Le Grice SF, Marino JP. A ribose sugar conformational switch in the LTR-retrotransposon Ty3 polypurine tract-containing RNA/DNA hybrid. *J Am Chem Soc* 2005;127:16344–5. doi:10.1021/ja0534203.

24. Turner KB, *et al.* SHAMS: combining chemical modification of RNA with mass spectrometry to examine polypurine tract-containing RNA/ DNA hybrids. *RNA* 2009;15:1605–13. doi:10.1261/rna.1615409.

25. Brinson RG, *et al.* Probing anomalous structural features in polypurine tract-containing RNA-DNA hybrids with neomycin B. *Biochemistry* 2009;48:6988–97. doi:10.1021/bi900357j.

26. Moran S, Ren RX, Kool ET. A thymidine triphosphate shape analog lacking Watson-Crick pairing ability is replicated with high sequence selectivity. *Proc Natl Acad Sci U S A* 1997;94:10506–11. doi:10.1073/ pnas.94.20.10506.

27. Moran S, Ren RX, Rumney S, Kool ET. Difluorotoluene, a nonpolar isostere for thymine, codes specifically and efficiently for adenine in

DNA replication. *J Am Chem Soc* 1997;119:2056–7. doi:10.1021/ ja963718g.

28. Rausch JW, Qu J, Yi-Brunozzi HY, Kool ET, Le Grice SF. Hydrolysis of RNA/DNA hybrids containing nonpolar pyrimidine isosteres defines regions essential for HIV type 1 polypurine tract selection. *Proc Natl Acad Sci U S A* 2003;100:11279–84. doi:10.1073/pnas.1932546100.

29. Lener D, Kvaratskhelia M, Le Grice SF. Nonpolar thymine isosteres in the Ty3 polypurine tract DNA template modulate processing and provide a model for its recognition by Ty3 reverse transcriptase. *J Biol Chem* 2003;278:26526–32. doi:10.1074/jbc.M302374200.

30. Jacobo-Molina A, *et al.* Crystal structure of human immunodeficiency virus type 1 reverse transcriptase complexed with double-stranded DNA at 3.0 A resolution shows bent DNA. *Proc Natl Acad Sci USA* 1993;90:6320–4. doi:10.1073/pnas.90.13.6320.

31. Nowak E, *et al.* Ty3 reverse transcriptase complexed with an RNA-DNA hybrid shows structural and functional asymmetry. *Nat Struct Mol Biol* 2014;21:389–96. doi:10.1038/nsmb.2785.

32. Wilhelm FX, Wilhelm M, Gabriel A. Extension and cleavage of the polypurine tract plus-strand primer by Ty1 reverse transcriptase. *J Biol Chem* 2003;278:47678–84. doi:10.1074/jbc.M305162200.

33. Wilhelm M, Uzun O, Mules EH, Gabriel A, Wilhelm FX. Polypurine tract formation by Ty1 RNase H. *J Biol Chem* 2001;276:47695–701. doi:10.1074/jbc.M106067200.

34. Abbondanzieri EA, *et al.* Dynamic binding orientations direct activity of HIV reverse transcriptase. *Nature* 2008;453:184–9. doi:10.1038/ nature06941.

35. Chung S, *et al.* Structure-activity analysis of vinylogous urea inhibitors of human immunodeficiency virus-encoded ribonuclease H. *Antimicrob Agents Chemother* 2010;54:3913–21. doi:10.1128/AAC.00434-10.

36. Liu S, Abbondanzieri EA, Rausch JW, Le Grice SF, Zhuang X. Slide into action: dynamic shuttling of HIV reverse transcriptase on nucleic acid substrates. *Science* 2008;322:1092–7. doi:10.1126/science.1163108.

37. Liu S, Harada BT, Miller JT, Le Grice SF, Zhuang X. Initiation complex dynamics direct the transitions between distinct phases of early HIV reverse transcription. *Nat Struct Mol Biol* 2010;17:1453–60. doi:10.1038/nsmb.1937.

38. Wilhelm M, Boutabout M, Wilhelm FX. Expression of an active form of recombinant Ty1 reverse transcriptase in Escherichia coli: a fusion protein containing the C-terminal region of the Ty1 integrase linked to the

reverse transcriptase-RNase H domain exhibits polymerase and RNase H activities. *Biochem J* 2000;348(Pt 2):337–42.

39. Wilhelm M, Heyman T, Boutabout M, Wilhelm FX. A sequence immediately upstream of the plus-strand primer is essential for plus-strand DNA synthesis of the Saccharomyces cerevisiae Ty1 retrotransposon. *Nucleic Acids Res* 1999;27:4547–52. doi:10.1093/nar/27.23.4547.

40. Bibillo A, Lener D, Tewari A, Le Grice SF. Interaction of the Ty3 reverse transcriptase thumb subdomain with template-primer. *J Biol Chem* 2005;280:30282–90. doi:10.1074/jbc.M502457200.

41. Brinson RG, Miller JT, Kahn JD, Le Grice SF, Marino JP. Applying thymine isostere 2,4-difluoro-5-methylbenzene as a NMR assignment tool and probe of homopyrimidine/homopurine tract structural dynamics. *Methods Enzymol* 2016;566:89–110. doi:10.1016/bs. mie.2015.05.009.

42. Dash C, Marino JP, Le Grice SF. Examining Ty3 polypurine tract structure and function by nucleoside analog interference. *J Biol Chem* 2006;281:2773–83. doi:10.1074/jbc.M510369200.

43. Nowak E, *et al.* Ty3 reverse transcriptase complexed with an RNA-DNA hybrid shows structural and functional asymmetry. Nat Struct Mol Biol 2014;21:389–96. doi:10.1038/nsmb.2785.

44. Bibillo A, Lener D, Klarmann GJ, Le Grice SF. Functional roles of carboxylate residues comprising the DNA polymerase active site triad of Ty3 reverse transcriptase. *Nucleic Acids Res* 2005;33:171–81. doi:10.1093/nar/gki150.

45. Sarafianos SG, *et al.* Structures of HIV-1 reverse transcriptase with pre- and post-translocation AZTMP-terminated DNA. *EMBO J* 2002;21:6614–24. doi:10.1093/emboj/cdf637.

46. Kaushik N, *et al.* Biochemical analysis of catalytically crucial aspartate mutants of human immunodeficiency virus type 1 reverse transcriptase. *Biochemistry* 1996;35:11536–46. doi:10.1021/bi960364x.

47. Saturno J, Lazaro JM, Blanco L, Salas M. Role of the first aspartate residue of the "YxDTDS" motif of phi29 DNA polymerase as a metal ligand during both TP-primed and DNA-primed DNA synthesis. *J Mol Biol* 1998;283:633–42. doi:10.1006/jmbi.1998.2121.

48. Uzun O, Gabriel A. A Ty1 reverse transcriptase active-site aspartate mutation blocks transposition but not polymerization. *J Virol* 2001;75:6337–47. doi:10.1128/JVI.75.14.6337-6347.2001.

49. Pandey M, Patel SS, Gabriel A. Kinetic pathway of pyrophosphorolysis by a retrotransposon reverse transcriptase. *PLoS One* 2008;3:e1389. doi:10.1371/journal.pone.0001389.

50. Yarrington RM, Chen J, Bolton EC, Boeke JD. Mn2+ suppressor mutations and biochemical communication between Ty1 reverse transcriptase and RNase H domains. *J Virol* 2007;81:9004–12. doi:10.1128/JVI.02502-06.

51. Rausch JW, *et al.* Interaction of p55 reverse transcriptase from the Saccharomyces cerevisiae retrotransposon Ty3 with conformationally distinct nucleic acid duplexes. *J Biol Chem* 2000;275:13879–87. doi:10.1074/jbc.275.18.13879.

52. Nowak E, *et al.* Structural analysis of monomeric retroviral reverse transcriptase in complex with an RNA/DNA hybrid. *Nucleic Acids Res* 2013;41:3874–87. doi:10.1093/nar/gkt053.

53. Lapkouski M, Tian L, Miller JT, Le Grice SFJ, Yang W. Complexes of HIV-1 RT, NNRTI and RNA/DNA hybrid reveal a structure compatible with RNA degradation. *Nat Struct Mol Biol* 2013;20:230–6. doi:10.1038/nsmb.2485.

54. Metzger W, Hermann T, Schatz O, Le Grice SF, Heumann H. Hydroxyl radical footprint analysis of human immunodeficiency virus reverse transcriptase-template.primer complexes. *Proc Natl Acad Sci USA* 1993;90:5909–13. doi:10.1073/pnas.90.13.5909.

55. Bebenek K, *et al.* A minor groove binding track in reverse transcriptase. *Nat Struct Biol* 1997;4:194–7. doi:10.1038/nsb0397-194.

56. Powell MD, *et al.* Residues in the alphaH and alphaI helices of the HIV-1 reverse transcriptase thumb subdomain required for the specificity of RNase H-catalyzed removal of the polypurine tract primer. *J Biol Chem* 1999;274:19885–93. doi:10.1074/jbc.274.28.19885.

57. Malik HS, Eickbush TH. Phylogenetic analysis of ribonuclease H domains suggests a late, chimeric origin of LTR retrotransposable elements and retroviruses. *Genome Res* 2001;11:1187–97. doi:10.1101/gr.185101.

58. Skalka AM. Retroviral DNA transposition: themes and variations. *Microbiol Spectr* 2014;2:MDNA300052014. doi:10.1128/microbiolspec.MDNA3-0005-2014.

59. Kirchner J, Sandmeyer SB. Ty3 integrase mutants defective in reverse transcription or 3′-end processing of extrachromosomal Ty3 DNA. *J Virol* 1996;70:4737–47.

60. Nymark-McMahon MH, Beliakova-Bethell NS, Darlix JL, Le Grice SF, Sandmeyer SB. Ty3 integrase is required for initiation of reverse transcription. *J Virol* 2002;76:2804–16. doi:10.1128/jvi.76.6.2804-2816.2002.

61. Nymark-McMahon MH, Sandmeyer SB. Mutations in nonconserved domains of Ty3 integrase affect multiple stages of the Ty3 life cycle. *J Virol* 1999;73:453–65.

62. Wilhelm FX, Wilhelm M, Gabriel A. Reverse transcriptase and integrase of the Saccharomyces cerevisiae Ty1 element. *Cytogenet Genome Res* 2005;110:269–87. doi:10.1159/000084960.

63. Wilhelm M, Wilhelm FX. Cooperation between reverse transcriptase and integrase during reverse transcription and formation of the preintegrative complex of Ty1. *Eukaryot Cell* 2006;5:1760–9. doi:10.1128/EC.00159-06.

64. Werner S, Wohrl BM. Soluble Rous sarcoma virus reverse transcriptases alpha, alphabeta, and beta purified from insect cells are processive DNA polymerases that lack an RNase H 3′ --> 5′ directed processing activity. *J Biol Chem* 1999;274:26329–36. doi:10.1074/jbc.274.37.26329.

65. Lauermann V, Boeke JD. The primer tRNA sequence is not inherited during Ty1 retrotransposition. *Proc Natl Acad Sci USA* 1994;91:9847–51. doi:10.1073/pnas.91.21.9847.

https://doi.org/10.1142/9789811249228_0003

Chapter 3

Experimental Systems for the Study of Non-LTR (LINE) Retrotransposons

Ivana Celic and Jeffrey S. Han*

1. Overview

The human genome is littered with millions of repetitive sequences, comprising more than half of the genome.[1,2] A select group of these sequences, called retrotransposons, is able to replicate by reverse transcribing their RNA into DNA at new genomic locations.[3] This process, called retrotransposition, continues to take place in our cells to this day. The currently active family of autonomous (self-replicating) elements in humans is called Long Interspersed Element-1 (LINE-1, or L1). L1 falls into the category of non-LTR retrotransposons, which replicate by reverse transcribing their RNAs in the nucleus, using primers derived from chromosome breaks. L1 is expressed at low levels in germ cells, and L1 is typically silenced in somatic cells.[4-6] However, certain abnormal states can lead to elevated L1 expression and L1 retrotransposition in both germ cells and somatic cells.

The potential importance of this aberrant L1 expression has become more apparent in recent years. Loss of transposon control coincides, in most cases, with germ cell demise and infertility in

*Department of Biochemistry and Molecular Biology, Tulane Cancer Center, Tulane University School of Medicine, New Orleans, LA 70112, United States.

model organisms.[6-13] This is also likely the case in humans, as alterations in key transposon control pathways are found in cases of human infertility.[14-18] In somatic cells, activation of L1 expression is associated with various neurological disorders, aging, and cancer.[19-29] Currently, we do not know the role, if any, that L1 plays in these disorders.

Questions regarding genetics, molecular and cell biology are often best answered using suitable experimental model systems/organisms. Each model system has advantages and disadvantages, and the best system to use depends on the particular question being posed. We believe that model systems will play a key role in helping to unravel a more detailed understanding of L1's mechanism, cell biology, and the interplay between transposons and disease states. To highlight the advantages of model systems, in this chapter, we will discuss how the unique properties of specific model systems contributed to seminal discoveries related to L1 biology, and how model systems may contribute to new discoveries moving forward.

2. Target-primed Reverse Transcription (R2 from *Bombyx mori* as a Model)

One of the most well-established steps of non-LTR replication is the initial cleavage of the target chromosome and subsequent reverse transcription of the RNA template, a process called target-primed reverse transcription (TPRT). During TPRT, an endonuclease encoded by the retrotransposon makes a single-stranded nick on the chromosome where integration will occur.[30] The 3′-hydroxyl DNA end generated by this nick is used to prime reverse transcription using the element's RNA as the template. TPRT was first described with *in vitro* biochemistry using the R2 element from the silkworm *Bombyx mori*. Several properties of R2 made this element ideal for the discovery of TPRT. First, while mammalian L1s encode two proteins critical for retrotransposition, R2 encodes only a single protein, the R2 ORF, greatly simplifying the reconstitution of a functional biochemical reaction. Second, R2 integrates into target DNA with restriction enzyme-like specificity. In contrast, L1 endonuclease has some promiscuity. Finally, unlike the apparent non-sequence-specific binding of L1 proteins, the R2 ORF has sequence preference for the R2 RNA it uses as a template. These stringent RNA/DNA sequence requirements of the R2 ORF allowed the design of RNA/DNA substrates that would give defined products, allowing unambiguous interpretation.

This clear and direct biochemical demonstration of TPRT has not been reproduced in other model systems due to the technical limitations described above. However, extensive sequencing analysis of existing and *de novo* post-integration events from a variety of organisms supports the TPRT model.[3,31–37] There are key differences between R2 ORF and the analogous L1 ORF2p.[38] R2 has a restriction enzyme-like endonuclease domain, while the L1 ORF2p endonuclease domain is homologous to apurinic/apyrimidinic endonucleases. The DNA-binding domains of R2 ORF and L1 ORF2p also appear unrelated. Thus, there may be mechanistic differences between R2 and L1 reverse transcription. Even so, the apparent conservation of TPRT suggests to us that basic principles for non-LTR retrotransposon replication will be conserved. Currently, the R2 system remains the only viable system for biochemically exploring mechanisms subsequent to non-LTR retrotransposon first-strand synthesis, such as second-strand cleavage and second-strand synthesis.

3. Infertility and Germ Cell Phenotypes (I Factor from *Drosophila Melanogaster* as a Model)

Transgenic overexpression of L1 allows detection of somatic retrotransposition events in mice, and aberrant expression of L1 during cancer or aging can lead to somatic retrotransposition events.[22–24,39–42] However, these somatic retrotransposition events are evolutionarily non-productive, as they are lost when the host organism dies. Evolutionarily successful transposons must be active in the germline. Consistent with this theory, endogenous transposons tend to be expressed in germ cells under normal conditions.[4–6] For example, in normal mouse testes, the detectable presence of L1 ribonucleoprotein particles (RNPs) is restricted primarily to prophase I of meiosis. L1 RNPs first appear in leptotene spermatocytes and persist until the pachytene stage.[6]

Although no one has yet cloned *de novo* retrotransposition events from spermatocytes, lines of evidence suggest that these are the biologically relevant cell types where L1 activity has evolved. The developmental stages of L1 RNP expression overlap with the appearance of germ-cell-specific small RNAs called Piwi-interacting RNAs (piRNAs).[43–47] Many of these piRNAs, complexed with their effector proteins (e.g., Mili and Miwi2 in mice), target transposable elements for repression via sequence homology, either by cleaving transposon RNA, or directing methylation of transposon DNA.[11,48] Mili binds piRNAs and exhibits slicer activity, cleaving complementary RNAs to produce more mature piRNAs through a ping-pong amplification loop.[49] Miwi2 binds mature piRNAs and localizes to the nucleus, suggesting a role in transcriptional silencing.[50,51] Knockout of Mili, Miwi2, or associated proteins (e.g., Maelstrom or Mov10l1) leads to elevated L1 expression, DNA damage, and meiotic arrest.[6,11–13] Recently, a transgenic allele of mouse L1 was developed that mimics expression patterns of endogenous L1s.[52] This allele has a marker that allows quantitation of retrotransposition. When piRNA biogenesis was disrupted, the normally low background of L1 retrotransposition increased 144-fold in leptotene/zygotene spermatocytes, the highest level found in the body.[52] The simplest explanation is that while L1 has evolved to retrotranspose during prophase I of meiosis, the host has evolved compensatory mechanisms which repress this L1 activity, leading to the observed low levels of retrotransposition in normal individuals.

While these recent findings regarding mammalian infertility and retrotransposons are remarkable, often lost in the shuffle is the fact that these findings were presaged decades ago by studies in the fruit fly *Drosophila melanogaster*. *Piwi*, the founding member of the Piwi gene family, was first identified in *Drosophila*, where *piwi* mutants exhibit germline failure along with activation of mobile elements.[53] Unlike humans, in whom genome content and retrotransposon activity have been dominated by L1 for millions of years, the *Drosophila* genome consists of a diverse population of active, low-copy transposable elements of both RNA and DNA varieties.[54,55] For some of these mobile elements, introduction into the *Drosophila melanogaster* genome was relatively recent, occurring in the wild but not in laboratory strains. Select crosses between wild and laboratory strains lead to a phenomenon called hybrid dysgenesis, characterized by genetic instability and infertility.[7,8] One of the best-known examples of this is I-R hybrid dysgenesis, in which a male "inducer", or I strain, is mated with a female "reactive", or R strain. The reciprocal cross does not produce dysgenic progeny. The genetic determinants responsible for the dysgenic phenotype were called I factors. We now know that I factor is similar to L1.[56]

Although mice with a defective piRNA pathway arrest during meiosis and are infertile, it is controversial whether the uncontrolled L1 expression is an underlying cause for infertility, because Piwi family members and piRNAs appear to have functions beyond transposon silencing. For example, postnatal disruption of *Mov10l1* allows relatively normal production of pre-pachytene piRNAs (which target transposons) and silencing of L1; however, pachytene piRNAs are lost and cells arrest at the round spermatid stage with elevated DNA damage.[57] In addition, some human *Piwi* mutations associated with human infertility show defects in late spermiogenesis independent of transposon reactivation.[58] These other contributing factors to germ cell arrest have made it difficult to tease out the importance of L1 activation in mammalian infertility. The classic genetic experiment of knocking out the candidate gene of interest is currently not practical with L1, which is estimated to have thousands of active copies in the mouse genome.[59] To complicate matters, the ubiquitous interspersed nature of L1 elements makes it difficult to interpret experiments designed to silence L1s by sequence homology. This is because there are pieces of L1 sequence in most mammalian genes,[60] and therefore globally silencing L1s would also silence other host genes. Finally, although using nucleoside inhibitors of reverse transcriptase (such as HIV inhibitors) can stop the retrotransposition reaction,[61] these inhibitors can have pleiotropic effects[62–65] and target a step which occurs *after* L1 cuts the chromosome. We believe DNA breaks may be responsible for L1 toxicity, and thus it is preferable to block L1 before DNA breaks occur.

While we do not yet have the technology to easily knock out all active L1s in the mouse, an analogous experiment has, in effect, already been done in *Drosophila*. The reactive R strains in I-R hybrid dysgenesis are strains that lack active copies of I factor. Without expression of active I factor mRNA, female R strains cannot produce effective piRNA silencing against I factor. Because these piRNA complexes are deposited into oocytes through the female germline,[66] female R strains produce eggs that are "defenseless" against I factor. Thus, the dysgenic I-R cross models the invasion of a new LINE retrotransposon in a naïve but otherwise wild-type host. This is an example of how the specific properties of a model organism can allow us to answer a question that may be intractable in mammals.

4. Host Factor Identification (*In Vitro Proteomics*)

The study of basic L1 biology currently has major limitations. We know that L1 needs to enter the nucleus in order to replicate. However, when expressed, a vast majority of L1 proteins are localized in the cytoplasm.[67–70] Much of the L1 proteins are in cytoplasmic granules, such as stress granules.[67,68] Evidence suggests that stress granules repress retrotransposition,[71–73] and much of what we look at is non-functional. This is consistent with the low absolute levels of retrotransposition even in cells expressing high amounts of L1.[3,31,74] The rare "active" L1 RNP is currently impossible to visually pinpoint among the sea of background. Thus, even some of the simplest questions in L1 cell biology remain unanswered, such as where the functional L1 RNPs are formed and how they traffic around the cell. The compact nature of a typical full-length LINE element belies the complexity of LINE retrotransposition, suggesting that LINE RNPs utilize host factors to assist in their replication. Given the inherent difficulty in studying the molecular/cell biology of L1, major efforts have been geared toward identifying L1/LINE host factors.[70,75,76] The hope is that by defining a complete L1 "interactome", we will have a better idea of how L1 replicates, and will be able to design the appropriate experiments to test these ideas. The two general strategies that have been used to identify host factors are biochemical and genetic.

Perhaps the easiest way to identify host factors involved with L1 retrotransposition is to purify complexes containing L1 proteins. Multiple labs have immunoprecipitated L1 ORF1p, the most abundant component of L1 RNPs, to look for interacting partners.[70,75,76] In each of these studies, epitope-labeled ORF1p, in the context of a full-length, active L1, was overexpressed in cultured cells (HeLa or 293T cells). ORF1p is a non-sequence-specific RNA-binding protein.[77] As expected, immunoprecipitation (IP) of ORF1p co-purifies predominantly RNA-binding proteins and RNA granules, most with unclear relevance to L1 biology. Roughly 90% of the candidate ORF1p interactors are lost in the presence of RNaseA, suggesting that many of these are non-specific interactions bridged by RNA.[75,76] Nevertheless, host factors that reduce L1 retrotransposition frequencies were identified, including the antiviral proteins MOV10, ZAP, and APOBEC family members.[70,75,76] Some general trends regarding ORF1p pulldowns are notable. In two large studies, out of a combined 88 potential host factors, 19 were restriction factors and zero were positive factors (i.e., factors that facilitate L1 retrotransposition).[75,76] Because a vast majority of L1 RNPs overexpressed in cultured cells appear cytoplasmically sequestered and are presumably not actively retrotransposing, it is not surprising that L1 RNPs isolated from cells are associated primarily with restriction factors. With high enough L1 expression, ORF2p pulldowns are also possible and have provided further insight into L1 RNP complexes.[70,78] Separate ORF1p/ORF2p and ORF2p-only complexes can be distinguished, with the ORF1/ORF2p complex likely to be identical to previously mentioned cytoplasmic ORF1p-containing complexes.[70,78] The ORF2p-only complex is postulated to be a nuclear complex, perhaps involved in TPRT. Enrichment of specific interacting partners in the context of L1 mutations known to effect target site cleavage and cDNA synthesis suggests an order in which they are assembled[78] — however, the functional role these putative L1 host factors play is unclear.

To sum, affinity proteomics has identified a wealth of host factors that interact with L1 RNPs and have the ability to alter retrotransposition frequency. As with all experimental systems, the purification of biochemical complexes has limitations. These methods are best suited to find the tightest and most abundant binders, and thus can only provide a glimpse into a subset of L1 host factors. Due to the large overlap in candidate host factors from three major studies in cultured cells, we expect that the "low hanging fruit" from these cell types has already been picked.[70,75,76] Further tweaking of *in vitro* conditions or analysis of how L1 mutants modify the composition of L1 RNPs complexes may provide further understanding into how known inter-actors contribute to retrotransposition. However, the discovery of novel host factors using biochemical purification will probably neces-sitate different starting material. The most obvious would be the interrogation of L1 RNPs in the germline. Although L1 expression levels in the germline are normally low, mice deficient in piRNA bio-genesis overexpress large amounts of endogenous L1 RNPs during meiosis. The L1 interactome in this cell population is a prime candi-date for biochemical characterization, and has the potential to uncover germ-specific L1 host factors.

5. Host Factor Identification (Genetics in Cell Culture and Yeast)

Because retrotransposons mobilize through an RNA intermediate, introns present in a retrotransposon should be spliced in *de novo* inser-tions.[79] Thus, placing an intron-interrupted reporter in L1 allows the monitoring of L1 retrotransposition events by various methods.[3,80,81] The development of these L1 retrotransposition assays has opened the door to perform genetic screens for L1 host factors in cultured cells. Two large-scale screens in human tissue culture cell lines have been performed. In one case, K562 or HeLa cells containing an inducible, neo[R]-marked L1 were transduced with a CRISPR sgRNA library.[82] The effect of sgRNAs on retrotransposition was determined by deep sequencing the sgRNA constructs from G418-selected populations of cells, and comparison to non-G418 selected cells. This CRISPR-based screen identified 164 restriction factors or positive factors. 89 of these factors showed concordant effects in K562 and HeLa cells, suggesting that the effects of the majority of the putative factors are not cell-line specific. The other large-scale screen in human cells was performed in HeLa-M2 cells containing an episomal GFP-marked L1.[83] These cells were plated in 384-well plates with arrayed siRNA pools to knock down individual human genes. Retrotransposition was monitored by microscopy by measuring the percentage of GFP-positive cells. This microscopy-based screen identified 220 putative L1 restriction host factors and 2681 putative L1-positive host factors. Remarkably, with so many genes identified, only five restriction factors and 22 positive factors were present in both screens, demonstrating that the setup of genetic screens can have a dramatic effect on the output.

A powerful model organism for genetics is the budding yeast *Saccharomyces cerevisae*. Although *S. cerevisiae* does not normally harbor non-LTR retrotransposons, an L1-like element called Zorro3 from the distantly related *Candida albicans* can be introduced into the *S. cerevisiae* genome and is able to retrotranspose much like a human L1.[84] This suggests that at minimum, the basal factors needed for LINE retrotransposition are conserved in budding yeast. This allowed the genetic screening of the yeast knockout collection for genes that effect Zorro3 retrotransposition in budding yeast, using an assay analogous to the human tissue culture assay.[85] Out of ~5000 strains assayed (representing knockout of all non-essential genes of yeast), 56 strains had a >90% decrease in retrotransposition. The most obvious pattern that emerged from the screen was that the endosomal sorting complex required for transport (ESCRT) is critical for LINE retrotransposition. Follow-up studies confirmed that this role for ESCRT is conserved with L1 in human cells. The initial yeast screen only identified 6 of the 17 genes essential for ESCRT function. Secondary screening revealed that all 17 of these ESCRT genes affected Zorro3 retrotransposition, suggesting that the original screen was not particularly sensitive and that alternative screen designs may uncover more host factors.

The beauty of genetics is that it makes no assumptions and therefore can uncover completely unexpected findings. However, many "hits" in genetic screens can also be indirect effects that do not play an integral role in the process of interest. With so many potential host factors identified in genetic screens, what is the next step to demonstrate the biological relevance to L1? In some cases, the path is straightforward — for example, HUSH and MORC2, identified in the CRISPR sgRNA screen, were found to bind the DNA of young L1 and promote histone H3 Lys9 trimethylation, leading to transcriptional silencing.[82] In cases involving post-transcriptional effects, the path forward is usually not as clear. The ESCRT complex has major effects on retrotransposition and there is strong evidence for direct physical interaction with L1.[85] A plausible function for ESCRT would be trafficking through membranes, presumably toward the nucleus, but this hypothesis is difficult to test experimentally. Likewise, reasonable models can be constructed for BRCA1 and DNA repair factors (found in both genetic screens in human cells) controlling L1 at the replication fork. However, direct evidence for these models is still lacking. The toolbox to delve deeper into the molecular and cell biology of L1 is somewhat limited — the development of more sensitive tools to visualize active L1 RNPs in the cytoplasm and nucleus is essential to truly take advantage of ever-increasing genetic data.

6. Moving Forward — The Role for Model Organisms

Much of what we know regarding L1 replication stems from experiments in cultured cells and analysis of existing and *de novo* L1 insertions. This has provided us with a useful framework for the basic requirements for non-LTR retrotransposition. We have a good idea of what domains/residues in L1 proteins are required for retrotransposition, and we know the range of insertion products that can be produced in various conditions. However, we must be cautious not to overgeneralize data obtained from cell culture systems. The overriding question that should be asked when interpreting data is as follows: "How does this fit within the context of organismal biology?" For example, APOBEC3 family members are antiviral cytidine deaminases first recognized for their inhibitory activities against HIV.[86,87] Based on the similarities between retroviruses and retrotransposons, multiple groups found that in cell culture, overexpressed APOBEC3s do indeed inhibit retrotransposons.[88-93] Does this mean that the normal function of APOBEC3s is to inhibit retrotransposons such as L1? Mice have a single APOBEC3 family member, and a knockout of APOBEC3 produces normally developed, fertile mice without predisposition to cancer or any observable disease.[94] APOBEC3 is also poorly expressed in the murine testes, suggesting that this protein does not normally play a role in retrotransposon defense. This underscores the importance of using model organisms to determine the biological relevance of experimental findings in other systems.

Another important aspect to consider when placing cell culture data in the appropriate biological context is whether there are differences between somatic retrotransposition and germline retrotransposition. Most experiments involving L1 use rapidly dividing tissue culture cells. This is a very different cellular environment than the presumed developmental stage (prophase I of meiosis) where L1 retrotransposition takes place. As suggested by recent data, L1 RNPs may enter the nucleus during mitosis when the nuclear envelope dissolves,[95] and retrotranspose during S phase.[36,37,83] However, in the male germline, the appearance of L1 RNPs occurs during leptonema through pachynema of prophase I.[6] This coincides with the appearance of piRNA complexes that restrict retrotransposons, suggesting that these stages of meiosis are an evolutionarily relevant "battleground" where L1 has evolved to retrotranspose. The developmental timing of *de novo* retrotransposition in MOV10l-deficient mice further supports retrotransposition at the onset of meiosis.[52] Interestingly, these meiotic stages occur after DNA replication and precede nuclear dissolution, suggesting that retrotransposition in the germline may require fundamentally different mechanisms for nuclear entry and chromatin interaction when compared to cancer cells.

To conclude, model organisms will play an essential role in evaluating the biological significance of the ever-expanding findings regarding sw. Ultimately, mouse will be the gold standard due to the similarity to humans and the many existing mouse models of cancer, aging, and neurologic disease. *Drosophila* provides many advantages over mouse, such as low propagation costs, fast generation time, easy genetic manipulation, a smaller genome, and an unparalleled collection of mutants. Like the mouse, models for cancer, aging, and brain disease have been developed in *Drosophila*. For these reasons, *Drosophila* should not be discounted, and in fact may be a preferred model organism, when examining fundamental *in vivo* responses to retrotransposition.

References

1. Lander ES, Linton LM, Birren B, *et al.* Initial sequencing and analysis of the human genome. *Nature* 2001;409(6822):860–921.
2. de Koning APJ, Gu W, Castoe TA, Batzer MA, Pollock DD. Repetitive elements may comprise over two-thirds of the human genome. *PLoS Genet* 2011;7(12):e1002384–12.
3. Moran JV, Holmes SE, Naas TP, DeBerardinis RJ, Boeke JD, Kazazian HH. High frequency retrotransposition in cultured mammalian cells. *Cell* 1996;87(5):917–27.
4. Branciforte D, Martin SL. Developmental and cell type specificity of LINE-1 expression in mouse testis: implications for transposition. *Mol Cell Biol* 1994;14(4):2584–92.
5. Trelogan SA, Martin SL. Tightly regulated, developmentally specific expression of the first open reading frame from LINE-1 during mouse embryogenesis. 1995;92(5):1520–4.
6. Soper SFC, van der Heijden GW, Hardiman TC, *et al.* Mouse maelstrom, a component of nuage, is essential for spermatogenesis and transposon repression in meiosis. *Dev Cell* 2008;15(2):285–97.
7. Picard G, L'Héritier P. A maternally inherited factor inducing sterility in Drosophila melanogaster. *Drosophila Inf Serv* 1971;46:54.
8. Kidwell MG, Kidwell JF, Sved JA. Hybrid dysgenesis in Drosophila melanogaster: a syndrome of aberrant traits including mutation, sterility and male recombination. *Genetics* 1977;86(4):813–33.
9. Ketting RF, Haverkamp TH, van Luenen HG, Plasterk RH. Mut-7 of C. elegans, required for transposon silencing and RNA interference, is a homolog of Werner syndrome helicase and RNaseD. *Cell* 1999;99(2):133–41.

10. Houwing S, Kamminga LM, Berezikov E, *et al.* A role for Piwi and piRNAs in germ cell maintenance and transposon silencing in Zebrafish. *Cell* 2007;129(1):69–82.

11. Kuramochi-Miyagawa S, Watanabe T, Gotoh K, *et al.* DNA methylation of retrotransposon genes is regulated by Piwi family members MILI and MIWI2 in murine fetal testes. *Genes Dev* 2008;22(7):908–17.

12. Carmell M, Vandekant H, Bourchis D, Bestor T, Derooij D, Hannon G. MIWI2 Is essential for spermatogenesis and repression of transposons in the mouse male germline. *Dev Cell* 2007;12(4):503–14.

13. Zheng K, Xiol J, Reuter M, *et al.* Mouse MOV10L1 associates with Piwi proteins and is an essential component of the Piwi-interacting RNA (piRNA) pathway. *Proc Natl Acad Sci* 2010;107(26):11841–6.

14. Gu A, Shi X, Long Y, Xia Y. Genetic variants in Piwi-interacting RNA pathway genes confer susceptibility to spermatogenic failure in a Chinese population. Hum Reprod 2010;25(12):2955–61.

15. Hadziselimovic F, Hadziselimovic N, Demougin P, Krey G, Oakeley E. Deficient expression of genes involved in the endogenous defense system against transposons in cryptorchid boys with impaired mini-ouberty. *Sex Dev* 2011;5(6):287–93.

16. Heyn H, Ferreira HJ, Bassas L, *et al.* Epigenetic disruption of the PIWI pathway in human spermatogenic disorders. *PLoS ONE* 2012;7(10): e47892–13.

17. Zhu X-B, Lu J-Q, Zhi E-L, *et al.* Association of a TDRD1 variant with spermatogenic failure susceptibility in the Han Chinese. *J Assist Reprod Genet* 2016;33(8):1–6.

18. Hadziselimovic F, Hadziselimovic NO, Demougin P, Krey G, Oakeley E. Piwi-pathway alteration induces LINE-1 transposon derepression and infertility development in cryptorchidism. *Sex Dev* 2015;9(2):98–104.

19. Bratthauer GL, Cardiff RD, Fanning TG. Expression of LINE-1 retrotransposons in human breast cancer. *Cancer* 1994;73(9):2333–6.

20. Harris CR, Normart R, Yang Q, *et al.* Association of nuclear localization of a long interspersed nuclear element-1 protein in breast tumors with poor prognostic outcomes. *Genes Cancer* 2010;1(2):115–24.

21. Rodić N, Sharma R, Sharma R, *et al.* Short communication. *Am J Pathol* 2014;1–8.

22. Lee E, Iskow R, Yang L, *et al.* Landscape of somatic retrotransposition in human cancers. *Science* 2012.

23. Iskow RC, Mccabe MT, Mills RE, *et al.* Natural mutagenesis of human genomes by endogenous retrotransposons. *Cell* 2010;141(7):1253–61.

24. Solyom S, Ewing AD, Rahrmann EP, *et al.* Extensive somatic L1 ret-
    rotransposition in colorectal tumors. *Genome Res* 2012.
25. De Cecco M, Criscione SW, Peterson AL, Neretti N, Sedivy JM,
    Kreiling JA. Transposable elements become active and mobile in the
    genomes of aging mammalian somatic tissues. 2013;5(12):867–83.
26. Cecco M, Ito T, Petrashen AP, *et al.* L1 drives IFN in senescent cells and
    promotes age-associated inflammation. *Nature* 2019;566(7742):1–33.
27. Simon M, Van Meter M, Ablaeva J, *et al.* LINE1 derepression in aged
    wild-type and SIRT6-deficient mice drives inflammation. *Cell Metab*
    2019;29(4):871–5.
28. Muotri AR, Chu VT, Marchetto MCN, Deng W, Moran JV, Gage FH.
    Somatic mosaicism in neuronal precursor cells mediated by L1 ret-
    rotransposition. *Nature* 2005;435(7044):903–10.
29. Muotri AR, Marchetto MCN, Coufal NG, Gage FH. L1 retrotransposition
    in neurons is modulated by MeCP2. *Nature* 2010;468(7322):443–6.
30. Luan DD, Korman MH, Jakubczak JL, Eickbush TH. Reverse transcrip-
    tion of R2Bm RNA is primed by a nick at the chromosomal target site:
    a mechanism for non-LTR retrotransposition. *Cell* 1993;72(4):
    595–605.
31. Feng Q, Moran JV, Kazazian HH, Boeke JD. Human L1 retrotranspo-
    son encodes a conserved endonuclease required for retrotransposition.
    Cell 1996;87(5):905–16.
32. Cost GJ, Feng Q, Jacquier A, Boeke JD. Human L1 element target-
    primed reverse transcription *in vitro.* 2002;21(21):5899–910.
33. Gilbert N, Lutz-Prigge S, Moran JV. Genomic deletions created upon
    LINE-1 retrotransposition. *Cell* 2002;110(3):315–25.
34. Symer DE, Connelly C, Szak ST, *et al.* Human l1 retrotransposition is
    associated with genetic instability in vivo. *Cell* 2002;110(3):327–38.
35. Gilbert N, Lutz S, Morrish TA, Moran JV. Multiple fates of L1 ret-
    rotransposition intermediates in cultured human cells. *Mol Cell Biol*
    2005;25(17):7780–95.
36. Flasch DA, Macia Á, Sanchez L, *et al.* Genome-wide de novo L1 ret-
    rotransposition connects endonuclease activity with replication. *Cell*
    2019;177(4):837–51.
37. Sultana T, van Essen D, Siol O, *et al.* The landscape of L1 retrotranspo-
    sons in the human genome is shaped by pre-insertion sequence biases
    and post-insertion selection. *Mol Cell* 2019;74(3):555–70.
38. Malik HS, Burke WD, Eickbush TH. The age and evolution of non-
    LTR retrotransposable elements. *Mol Biol Evol* 1999;16(6):793–805.

39. An W, Han JS, Wheelan SJ, *et al.* Active retrotransposition by a synthetic L1 element in mice. *Proc Natl Acad Sci* 2006;103(49):18662–7.

40. Babushok DV, Ostertag EM, Courtney CE, Choi JM, Kazazian HH. L1 integration in a transgenic mouse model. *Genome Res* 2006;16(2): 240–50.

41. Kano H, Godoy I, Courtney C, *et al.* L1 retrotransposition occurs mainly in embryogenesis and creates somatic mosaicism. *Genes Dev* 2009;23(11):1303–12.

42. O'Donnell KA, An W, Schrum CT, Wheelan SJ, Boeke JD. Controlled insertional mutagenesis using a LINE-1 (ORFeus) gene-trap mouse model. *Proc Natl Acad Sci* 2013;110(29):E2706–13.

43. Grivna ST, Beyret E, Lin H. A novel class of small RNAs in mouse spermatogenic cells. *Genes Dev* 2006;20(13):1709–14.

44. Aravin A, Gaidatzis D, Pfeffer S, *et al.* A novel class of small RNAs bind to MILI protein in mouse testes. *Nature* 2006;442(7099):203–7.

45. Girard A, Sachidanandam R, Hannon GJ, Carmell MA. A germline-specific class of small RNAs binds mammalian Piwi proteins. *Nature* 2006;442(7099):199–202.

46. Watanabe T, Takeda A, Tsukiyama T, *et al.* Identification and characterization of two novel classes of small RNAs in the mouse germline: retrotransposon-derived siRNAs in oocytes and germline small RNAs in testes. *Genes Dev* 2006;20(13):1732–43.

47. Lau NC, Seto AG, Kim J, *et al.* Characterization of the piRNA complex from rat testes. *Science* 2006;313(5785):363–7.

48. Aravin AA, Sachidanandam R, Bourc'his D, *et al.* A piRNA pathway primed by individual transposons is linked to de novo DNA methylation in mice. *Mol Cell* 2008;31(6):785–99.

49. De Fazio S, Bartonicek N, Di Giacomo M, *et al.* The endonuclease activity of Mili fuels piRNA amplification that silences LINE1 elements. *Nature* 2011;480(7376):259–63.

50. Aravin AA, Sachidanandam R, Bourc'his D, *et al.* A piRNA pathway primed by individual transposons is linked to de novo DNA methylation in mice. *Mol Cell* 2008;31(6):785–99.

51. Reuter M, Chuma S, Tanaka T, Franz T, Stark A, Pillai RS. Loss of the Mili-interacting Tudor domain-containing protein-1 activates transposons and alters the Mili-associated small RNA profile. *Nat Struct Mol Biol* 2009;16(6):639–46.

52. Newkirk SJ, Lee S, Grandi FC, *et al.* Intact piRNA pathway prevents L1 mobilization in male meiosis. *Proc Natl Acad Sci USA* 2017;114(28): E5635–44.

53. Lin H, Spradling AC. A novel group of pumilio mutations affects the asymmetric division of germline stem cells in the Drosophila ovary. *Development* 1997;124(12):2463–76.

54. Adams MD, Celniker SE, Holt RA, *et al.* The genome sequence of Drosophila melanogaster. *Science* 2000;287(5461):2185–95.

55. McCullers TJ, Steiniger M. Transposable elements in Drosophila. *Mob Genet Elements* 2017;7(3):1–18.

56. Fawcett DH, Lister CK, Kellett E, Finnegan DJ. Transposable elements controlling I-R hybrid dysgenesis in D. melanogaster are similar to mammalian LINEs. *Cell* 1986;47(6):1007–15.

57. Zheng K, Wang PJ. Blockade of pachytene piRNA biogenesis reveals a novel requirement for maintaining post-meiotic germline genome integrity. *PLoS Genet* 2012;8(11):e1003038.

58. Gou L-T, Kang J-Y, Dai P, *et al.* Ubiquitination-deficient mutations in human Piwi cause male infertility by impairing histone-to-protamine exchange during spermiogenesis. *Cell* 2017;169(6):1090–1095.e13.

59. Goodier JL, Ostertag EM, Du K, Kazazian HH. A novel active L1 retrotransposon subfamily in the mouse. *Genome Res* 2001;11(10):1677–85.

60. Han JS, Szak ST, Boeke JD. Transcriptional disruption by the L1 retrotransposon and implications for mammalian transcriptomes. *Nature* 2004;429(6989):268–74.

61. Dai L, Huang Q, Boeke JD. Effect of reverse transcriptase inhibitors on LINE-1 and Ty1 reverse transcriptase activities and on LINE-1 retrotransposition. *BMC Biochem* 2011;12(1):18.

62. Fang J-L, McGarrity LJ, Beland FA. Interference of cell cycle progression by zidovudine and lamivudine in NIH 3T3 cells. *Mutagenesis* 2009;24(2):133–41.

63. Olivero OA. Zidovudine induces S-phase arrest and cell cycle gene expression changes in human cells. *Mutagenesis* 2005;20(2):139–46.

64. Roskrow M, Wickramasinghe SN. Acute effects of 3'-azido-3'-deoxythymidine on the cell cycle of HL60 cells. *Clin Lab Haematol* 1990;12(2):177–84.

65. Fowler BJ, Gelfand BD, Kim Y, *et al.* Nucleoside reverse transcriptase inhibitors possess intrinsic anti-inflammatory activity. *Science* 2014;346(6212):1000–3.

66. Brennecke J, Malone CD, Aravin AA, Sachidanandam R, Stark A, Hannon GJ. An epigenetic role for maternally inherited piRNAs in transposon silencing. *Science* 2008;322(5906):1387–92.

67. Goodier JL, Zhang L, Vetter MR, Kazazian HH. LINE-1 ORF1 protein localizes in stress granules with other RNA-binding proteins, including components of RNA interference RNA-induced silencing complex. *Mol Cell Biol* 2007;27(18):6469–83.
68. Doucet AJ, Hulme AE, Sahinovic E, *et al.* Characterization of LINE-1 ribonucleoprotein particles. *PLoS Genet* 2010;6(10).
69. Goodier JL, Mandal PK, Zhang L, Kazazian HH. Discrete subcellular partitioning of human retrotransposon RNAs despite a common mechanism of genome insertion. *Hum Mol Genet* 2010.
70. Taylor MS, Lacava J, Mita P, *et al.* Affinity proteomics reveals human host factors implicated in discrete stages of LINE-1 retrotransposition. *Cell* 2013;155(5):1034–48.
71. Guo H, Chitiprolu M, Gagnon D, *et al.* Autophagy supports genomic stability by degrading retrotransposon RNA. *Nat Commun* 2014; 5:1–11.
72. Goodier JL, Cheung LE, Kazazian HH. MOV10 RNA helicase is a potent inhibitor of retrotransposition in cells. *PLoS Genet* 2012;8(10):e1002941.
73. Hu S, Li J, Xu F, *et al.* SAMHD1 inhibits LINE-1 retrotransposition by promoting stress granule formation. *PLoS Genet* 2015;11(7):e1005367–27.
74. Han JS, Boeke JD. A highly active synthetic mammalian retrotransposon. *Nature* 2004;429(6989):314–8.
75. Goodier JL, Cheung LE, Kazazian HH. Mapping the LINE1 ORF1 protein interactome reveals associated inhibitors of human retrotransposition. *Nucleic Acids Res* 2013;41(15):7401–19.
76. Moldovan JB, Moran JV. The zinc-finger antiviral protein ZAP inhibits LINE and Alu retrotransposition. *PLoS Genet* 2015;11(5):e1005121–34.
77. Kolosha VO, Martin SL. High-affinity, non-sequence-specific RNA binding by the open reading frame 1 (ORF1) protein from long interspersed nuclear element 1 (LINE-1). *J Biol Chem* 2003;278(10):8112–7.
78. Taylor MS, Altukhov I, Molloy KR, *et al.* Dissection of affinity captured LINE-1 macromolecular complexes. *Elife* 2018;7:e30094.
79. Boeke JD, Garfinkel DJ, Styles CA, Fink GR. Ty elements transpose through an RNA intermediate. *Cell* 1985;40(3):491–500.
80. Ostertag EM, Prak ET, DeBerardinis RJ, Moran JV, Kazazian HH. Determination of L1 retrotransposition kinetics in cultured cells. *Nucleic Acids Res* 2000;28(6):1418–23.

81. Xie Y, Rosser JM, Thompson TL, Boeke JD, An W. Characterization of L1 retrotransposition with high-throughput dual-luciferase assays. *Nucleic Acids Res* 2011;39(3):e16–6.
82. Liu N, Lee CH, Swigut T, *et al*. Selective silencing of euchromatic L1s revealed by genome-wide screens for L1 regulators. *Nature* 2018;553(7687):228–32.
83. Mita P, Sun X, Fenyö D, *et al*. BRCA1 and S phase DNA repair pathways restrict LINE-1 retrotransposition in human cells. *Nat Struct Mol Biol* 2020;27(2):1–44.
84. Dong C, Poulter RT, Han JS. LINE-like retrotransposition in Saccharomyces cerevisiae. *Genetics* 2009;181(1):301–11.
85. Horn AV, Celic I, Dong C, Martirosyan I, Han JS. A conserved role for the ESCRT membrane budding complex in LINE retrotransposition. *PLoS Genet* 2017;13(6):e1006837–28.
86. Mangeat B, Turelli P, Caron G, Friedli M, Perrin L, Trono D. Broad antiretroviral defence by human APOBEC3G through lethal editing of nascent reverse transcripts. *Nature* 2003;424(6944):99–103.
87. Zhang H, Yang B, Pomerantz RJ, Zhang C, Arunachalam SC, Gao L. The cytidine deaminase CEM15 induces hypermutation in newly synthesized HIV-1 DNA. *Nature* 2003;424(6944):94–8.
88. Dutko JA, Schäfer A, Kenny AE, Cullen BR, Curcio MJ. Inhibition of a yeast LTR retrotransposon by human APOBEC3 cytidine deaminases. *Curr Biol* 2005;15(7):661–6.
89. Schumacher AJ, Nissley DV, Harris RS. APOBEC3G hypermutates genomic DNA and inhibits Ty1 retrotransposition in yeast. *Proc Natl Acad Sci* USA 2005;102(28):9854–9.
90. Bogerd HP, Wiegand HL, Hulme AE, *et al*. Cellular inhibitors of long interspersed element 1 and Alu retrotransposition. 2006;103(23): 8780–5.
91. Muckenfuss H, Held U, Perkovic M, *et al*. APOBEC3 proteins inhibit human LINE-1 retrotransposition. *J Biol Chem* 2006;281(31): 22161–72.
92. Stenglein MD, Harris RS. APOBEC3B and APOBEC3F inhibit L1 retrotransposition by a DNA deamination-independent mechanism. *J Biol Chem* 2006;281(25):16837–41.
93. Kinomoto M, Kanno T, Shimura M, *et al*. All APOBEC3 family proteins differentially inhibit LINE-1 retrotransposition. *Nucleic Acids Res* 2007;35(9):2955–64.

94. Mikl MC, Watt IN, Lu M, *et al.* Mice deficient in APOBEC2 and APOBEC3. *Mol Cell Biol* 2005;25(16):7270–7.
95. Mita P, Wudzinska A, Sun X, *et al.* LINE-1 protein localization and functional dynamics during the cell cycle. *Elife* 2018;7:e30058.

https://doi.org/10.1142/9789811249228_0004

Chapter 4
Alu Elements and Human Disease

Hanlin Yang, Maria E. Morales, and Astrid M. Roy-Engel*

1. Transposable Elements

A transposable element (TE) is a mobile DNA sequence that has the ability to change its position in a genome and can create recognizable transposon-derived repeats. Because transposition entails the duplication of genetic material, the definition of TE also includes genetic sequences that do not mobilize themselves, but are still able to generate copies that insert elsewhere in the genome. Such repetitive sequences were initially estimated to contribute close to 50% of the human genome,[1] but further analyzes using highly sensitive strategies indicate that TEs contribute up to two thirds of the genome.[2] Through millions of years of evolution, TE insertions have significantly contributed to genetic diversity.

Overall, transposable elements fall into two classes (Class I and Class II) based on their mechanism of insertion (see Fig. 1).[3]

Both classes of TEs can be found across kingdoms. However, in this chapter, we will focus on those elements present in the human genome. Figure 2 illustrates the basic classification of TEs found in the human genome based on the type of mobilization utilized.

---

*Tulane Cancer Center SL-66, Department of Epidemiology, Tulane University Health Sciences Center, 1430 Tulane Ave., New Orleans, LA 70112, USA.

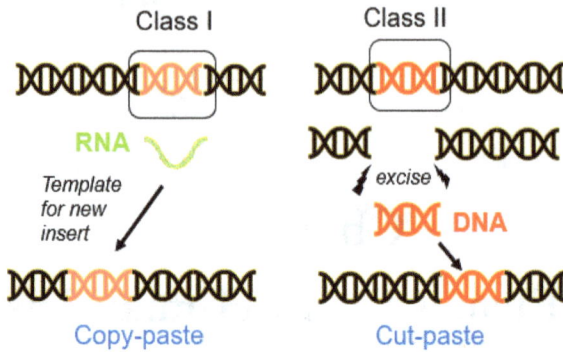

Figure 1. Mobile elements are divided into Class I (copy and paste) and Class II (cut and paste) based on the mechanism of amplification via an RNA or DNA intermediate (for class I, it would be important to show that the original mobile element remains after the new insert).

Class II TEs amplify via DNA intermediates for amplification in a "cut and paste" mechanism. Different groups of DNA transposons are further subdivided based on their mode of replication.[4] For mobilization, autonomous DNA elements encode a transposase that binds to the short terminal inverted repeats (TIRs) located within the transposon to catalyze both the DNA cleavage and strand transfer steps of the transposition reaction. The transposase usually recognizes a small target sequence where it introduces the transposon element.[5,6] DNA transposons make up ~3% of the human genome.[1] However, this type of TE is inactive in the human genome and the copies represent fossil remnants of dead elements.

Class I TEs, also known as retrotransposons, amplify via an RNA intermediate by a "copy and paste" mechanism in a process referred to as retrotransposition.[7] In the human genome, the vast majority of retrotransposed sequences commonly display three characteristic hallmarks: (1) flanking target site duplications (TSD) of varying lengths, (2) the usual presence of a 3' poly adenosine stretch or A-tail (with the exception of the tailless retropseudogenes[8]), and (3) usual integration occurring at an AT-rich sequence (some exceptions). Class I TEs are further subclassified, based on the presence or absence of a 100–300-bp LTR flanking the element, into LTR retrotransposons and non-LTR retrotransposons.

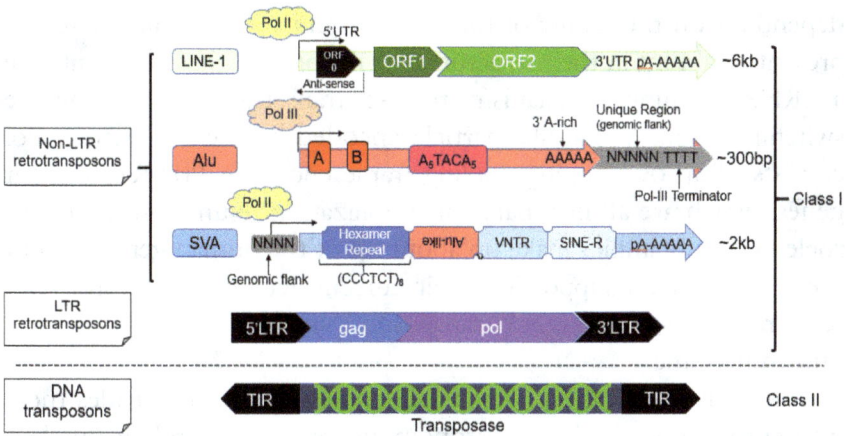

Figure 2. Schematic diagram of the basic structural organization of Class I and Class II elements present in the human genome. The **non-LTR elements** include long interspersed nuclear elements (e.g., LINE-1 or L1), short interspersed nuclear elements (e.g., SINES such as Alu and MIR, not shown), and the retroposon SVA. Full-length **LINE-1** elements are about 6 kb in length and are composed of a 5'UTR that contains an internal RNA polymerase II promoter, includes three open reading frames (ORF0, ORF1, and ORF2) and a 3'UTR that contains a polyadenylation signal (pA) and an A-tail. **Alu** elements have two non-identical sequences separated by a middle adenine-rich region (A_5TACA_5) and flanked at the 3' end by an A-rich region usually referred to as an "'A-tail". However, please note that Alu transcripts will optionally contain the 3' flanking unique genomic sequence (shown in gray) past the A-tail due to the lack of an RNA pol III terminator sequence (TTTT) within the Alu sequence. The Alu left monomer contains the internal bipartite RNA pol III promoter with an A and B box. **SVA** elements are composite elements that contain a 5' region with a variable number of CCCTCT repeats, an antisense "Alu-like sequence", a variable number tandem repeat (VNTR) region, and an HERVK-like region referred to as SINE-R. SVA elements also contain a 3' A-tail. **Long Terminal Repeat (LTR) elements**, sometimes referred to as human endogenous retroviruses (HERVs), structurally resemble retroviruses and contain two LTR sequences flanking the encoded polyproteins capsid (gag), protease, polymerase (pol), and possibly envelope (not shown). Class II **DNA transposons** are flanked by terminal inverted repeats (TIR) and encode a transposase that excises the transposon out of the donor position and reintegrates it into a new location. Note: the diagram is not drawn to scale.

Autonomous LTR retrotransposons utilize an RNA-mediated transposition with encoded protein factors gag and pol, flanked by 300-bp-long terminal repeats.[9] These elements resemble retroviruses in their structure.[9] Replication of these elements shows some variation depending on the family of the element and which components are present or absent in their structure.[10] A majority of these elements use a tRNA priming mechanism reverse transcription and template switching using a viral-like particle encoded by the *gag* gene.[11] Pol consists of an overlapping reading frame encoding Prt, RT, and Int genes, which are all necessary for mobilization. Numerous LTR retroelement subfamilies make up about 8% of the human genome,[1] but there are no data supporting their current activity. Therefore, LTR retrotransposons, like DNA transposons, are deemed inactive in the human genome.

Further classification of non-LTR retrotransposons divides them into autonomous and non-autonomous groups depending on their ability to provide the machinery for self-mobilization. Three types of non-LTR retrotransposons represent nearly all human retrotransposons: (1) the autonomous Long Interspersed Elements (LINEs), the non-autonomous (2) Short Interspersed Elements (SINEs), and (3) SINE-VNTR-Alus, (SVAs). For mobilization, the non-autonomous retrotransposons parasitize the factors generated by LINEs. Although not classified as retrotransposons, processed pseudogenes or retropseudogenes also utilize LINE factors for their creation.[12] By mass, LINEs represent the most predominant non-LTR retrotransposons, accounting for 21% of the human genome, of which L1 elements account for 17%.[1]

L1, an autonomous and currently active non-LTR retrotransposon in the human genome, utilizes an internal RNA polymerase II promoter[13] to transcribe a bicistronic RNA[14] that encodes for ORF0 of unknown function.[15] ORF 1, a nucleic acid chaperone[16] and ORF2 that contains a functional endonuclease and reverse transcriptase domain.[17,18] There are ~520,000 genomic copies of L1 in the human genome,[1] but there are a little over 5000 full-length 6-kb L1 elements.[19] A majority of other copies consist of 5′ truncated elements with an average length of 900 bp and a few disrupted by other insertions elements.[19] A majority of the full-length L1 copies contain inactivating point mutations in their open reading frame, resulting in an estimate of ~100 active L1 elements capable of retrotransposition,[19] many of which are likely polymorphic between individuals.[20] However, these active elements show significant differences in retrotransposition rate, where only a small portion contributes to the bulk of activity and is considered a "hot element".[19] L1 is considered the main driver of non-autonomous element amplification in the human genome.

The non-autonomous non-LTR retrotransposons fall into the general class referred to as retroposons. Of these, SINEs contribute about 13% of the human genome, of which Alu constitutes 11% of genome mass.[1] In general, SINEs are grouped into three superfamilies called CORE-SINE, V-SINE, and AmnSINE.[21] CORE-SINEs share a common 5′ sequence but vary in their 3′ends. V-SINES share a central sequence with varying 3′ sequence. In contrast, AmnSINEs are chimeric, derived from 5S rRNA and tRNA genes. Overall, SINEs share two basic characteristics: (1) short length (~100–300 bp) and (2) they are derived from RNA polymerase III (RNA pol III) transcribed RNAs.[22] Active SINEs retain the internal RNA pol III A and B boxes, which generate the RNA intermediate used in their amplification process. Most SINEs are derived from tRNAs (e.g., B2 and ID in rodents) with only two (primate Alu and the rodent B1) reported to derive from 7SL, the RNA component of the signal recognition particle (SRP) involved in protein secretion.[23] There are a few examples of SINEs derived from other RNA pol III genes, such as the 5S gene. In some cases, one can find composite SINEs which are derived from two different genes (part 7SL, part tRNA). In the human genome, with over one million copies, Alu is by far the most abundant (by copy number) of all the TEs present in the human genome.

2. "Aluology"

2.1 *Alu evolution*

Alu elements are considered as a "dimeric" sequence that was derived from two non-identical "monomers" of the "Alu domain" of the 7SL RNA gene,[24] the nucleic acid component of the SRP ribonucleoprotein.[7] These two regions or monomers connect via an A-rich stretch to form the Alu element of approximately 300 bp in length. At the 3′

end, the Alu transcript contains a run of As referred to as an "A-tail" even if it is not always the actual end of the Alu transcript. Because the Alu sequence lacks an RNA Pol III terminator signal, transcription continues downstream of the "A-tail" in the genomic flanking sequence until it encounters four Ts or more. Thus, transcripts from different Alu loci will have different 3′ sequences, referred to as the "unique region" (see Alu diagram in Fig. 2). However, not all Alu transcripts contain a unique region, as RNA Pol III terminators may locate immediately downstream of the 3′A-rich region. In some instances, sequencing of the unique region of the Alu RNA could provide sufficient information to identify the individual locus that generated the transcript.[25]

Two separate phases define Alu evolution: a precursor monomeric phase and a dimeric phase of the sequence we currently define as Alu. The insertion of a partial deletion of a pseudogene of the 7SL RNA gene likely gave rise to the Alu monomeric progenitor.[7] One model ("the SINE genesis model") proposes that SINEs arose from Pol-III defective transcript fragments that "escaped" degradation by the exosome in the nucleus.[26] In this model, the Trf4/Air2/Mtr4 polyadenylation (TRAMP) complex tags the Pol III transcript fragment by adding a 3′ poly(A) for degradation (Fig. 3). In the case of the Alu, instead of undergoing degradation, the 3′ poly(A)-tagged 7SL RNA fragment undergoes retrotransposition to generate the ancestral precursor. Initial evolutionary analyzes estimated the appearance of this monomeric progenitor sequence about 65 million years ago (mya),[27,28] but it could have occurred even earlier, sometime before the primate/rodent evolutionary divergence. Because the progenitor sequence retained the internal polymerase III (pol III) promoter of the 7SL gene, it retained the potential to transcribe RNA. Subsequently, transcripts from this progenitor likely provided the RNA template to generate more retrotransposed copies in the genome. This first fossil Alu monomer (FAM) family possibly arose early in the mammalian radiation.[29]

Amplification and subsequent sequence divergence of the FAM family generated the free left Alu monomer (FLAM) and the free right Alu monomer (FRAM) families of monomeric elements (Fig. 3).[30]

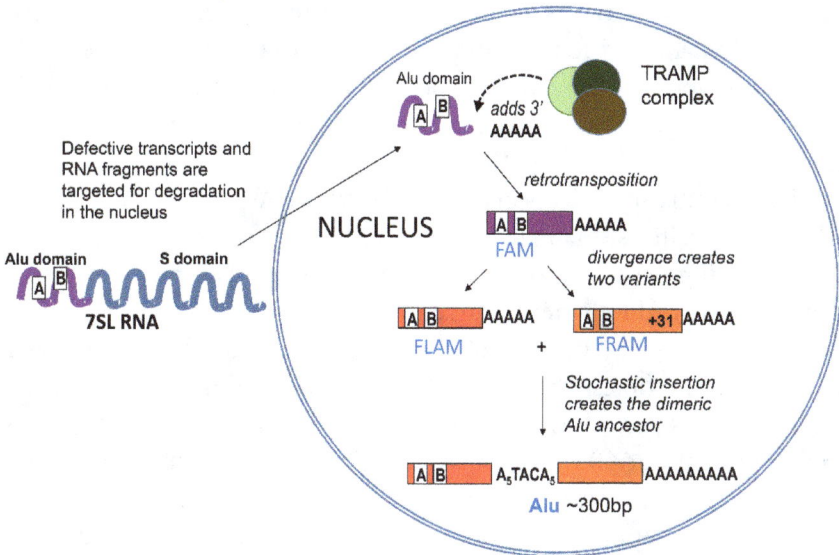

Figure 3. Proposed model of genesis of Alu SINE. Possibly a partially degraded or defective 7SL transcript was targeted for destruction in the nucleus by the exosome and in the process it was polyadenylated by the TRAMP complex. This first transcript likely gave rise to the ancestral FAM and that subsequently evolved into the ancestral dimeric Alu sequence.

FLAM lacks 31 bp present in FRAM. The presence of human-specific Alu monomers indicates that some copies remained active after the emergence of humans.[31] Proposed models suggest that stochastic insertion next to each other led to the creation of a FLAM–FRAM fusion that gave rise to the dimeric progenitor of the Alu family.[30] Sometime during this phase, the A and B boxes of the FRAM-derived region of the dimeric Alu diverged, losing their ability to support Pol-III transcription.

The dimeric phase of Alu expansion started about 65 million years ago (mya), peaking between 60 and 35 mya to decline to the current levels of activity.[32] The current estimate of Alu retrotransposition rate posits one new insertion every approximately 20 births,[33] which represents approximately two orders of magnitude lower than the peak rate. Alu amplification started early in primate evolution contributing to the accumulation of over 1 million copies present in the human genome. The lack of a specific mechanism for the removal of retro-transposons allowed for the highly effective accumulation of Alu elements. However, some Alu sequences can be lost through non-specific mechanisms (genomic deletions, non-allelic homologous recombination, etc.).

During this amplification period, the active Alu elements diverged, whereby specific subfamilies of Alu sequences contributed to new inserts during specific evolutionary time periods. The majority of the 1 million plus Alu elements currently present in the human genome occurred during the peak amplification period ~55 mya contributing to ~850,000 copies of the AluJ and AluSx subfamilies (Fig. 4). Following this period, the rate of amplification slowed with the accumulation of about 40,000 copies of the Alu Sg1 subfamily. This group of subfamilies (all AluJs and AluSs) is collectively referred to as the "old" Alu elements. While some individual old Alu loci may be able

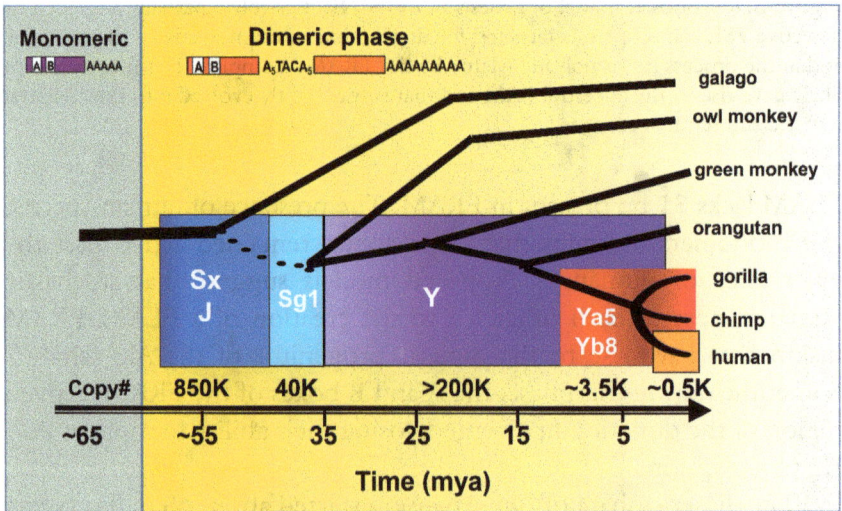

Figure 4. Evolutionary amplification of Alu elements. The dimeric phase of Alu amplification started about 65 million years ago. The old Alu subfamilies J, Sx, and Sg1 were most active around 35 to 55 million years ago accumulating over 850,000 copies representing the majority of the Alu elements present today in the human genome. The amplification rate of Alu decreased with evolutionary time as observed by the reduction of the copy numbers. Currently, the young Alu subfamilies (Y, Ya5, Yb8, Yc1, etc.) contribute to all the known polymorphisms in the human genome.

to generate transcripts,[34] they are currently considered as retrotranspositionally inactive. Although there are some reports of polymorphic old Alu inserts, these are rare and unlikely the result of a retrotransposition event. Instead, they may represent a deletion event.[35] About 35 mya, the "young Alu" subfamilies started amplifying, with the Alu Y subfamily and its subsequent younger Alu subfamily members such as Ya5, Ya8, and Yb8 continuing into modern times (Fig. 4). Many of these younger Alu subfamilies inserted after the human radiation, making them human specific. Thus, the young Alu Ya and Yb subfamilies and their variants most likely account for all human-specific insertions.[36,37] Because the active young subfamilies continue to insert in human genomes, it creates a variation between different humans. If two humans have the exact same polymorphic Alu present in their genome, it indicates the likelihood of a shared common ancestor. Thus, relationships between people from different regions or different families can be traced. This feature, allows polymorphic Alu elements to serve both as a forensic tool and a population marker.[38]

A vast majority of Alu copies in the human genome are unable to amplify, and only a few retrocompetent Alu loci referred to as "source elements" contribute to new inserts. This concept was first introduced as the "master" element model.[39] Throughout evolution, the source elements lose functionality through the accumulation of mutations at a neutral rate, and older copies accrue more sequence degradation than newer copies. Thus, a particular source or group of source elements will dominate during a given period of time. However, the appearance of short-lived hyperactive copies, referred to as "stealth drivers", also influenced Alu amplification dynamics.[40] Currently, these young Alu subfamilies are the only ones active in the human genome and create genetic polymorphism between individuals.[41]

2.2 *Alu retrotransposition*

Alu elements are devoid of coding sequence and unable to generate the proteins needed for retrotransposition. Thus, Alu depends entirely on the L1 element machinery for its own mobilization.[42] Although

L1 requires both the ORF1p and the ORF2p proteins for retrotransposition, Alu only strictly requires ORF2p.[42,43]

After the Alu RNA is transcribed by RNA polymerase III, several proteins bind to the RNA to form an RNA–protein complex (RNP). Being derived from 7SL, the Alu transcript retains a similar secondary structure that could allow it to form an RNP with the SRP9 and SRP14 proteins that normally bind the "Alu domain" of 7SL (Figs. 5(A) and 5(B)). These proteins play an important role in targeting the SRP to the ribosome. In particular, SRP14 is critical to maintain the SRP's ability to induce elongation arrest.[44] Similar to how SRP9 and SRP14 help target the SRP RNP,[45] these proteins are proposed to also help target Alu RNP to ribosomes.[46] Furthermore, the binding of the Alu RNP to ribosomes increases it chance to encounter a translating L1 and sequester the factors needed (e.g., ORF2) for its own retrotransposition (Fig. 5(C)). A previous study demonstrated that mutations in the Alu sequences altering theSRP9/14-Alu RNA interaction hindered the retrotransposition capability of Alu, supporting a role for these proteins in the Alu amplification mechanism.[47,48] In addition, the connection of both Alu and L1 sharing a 3′ end A-tail was suggested as a potential mechanism for Alu to be able to hijack the L1 machinery.[49] Later studies further supported this connection, as the cytoplasmic polyA-binding protein (PABP) interacts with the poly A region of SINEs and is thought to help target the Alu RNP to the ribosomes.[26,50,51] Interestingly, the PABP has been shown to play a role in L1 retrotransposition.[52] However, the role of PABP in SINE retrotransposition is unclear. The model suggests that these proteins, by promoting the proximity of Alu RNP to ribosomes translating the L1 mRNA, likely enable Alu to hijack the newly translated L1 proteins.[42]

How the Alu RNA returns to the nucleus for the insertion process is unclear. Alu insertion is thought to occur via the target-primed reverse transcription (TPRT) model used by the R2 element in *Bombyx mori*[53] (Fig. 6), as Alu inserts show the typical hallmarks of inserts that undergo this process. In this model, the endonuclease activity of ORF2p from L1 cleaves at an A/T-rich consensus sequence (5′-TTTT/AA-3) at the target site. The T-rich sequence base pairs

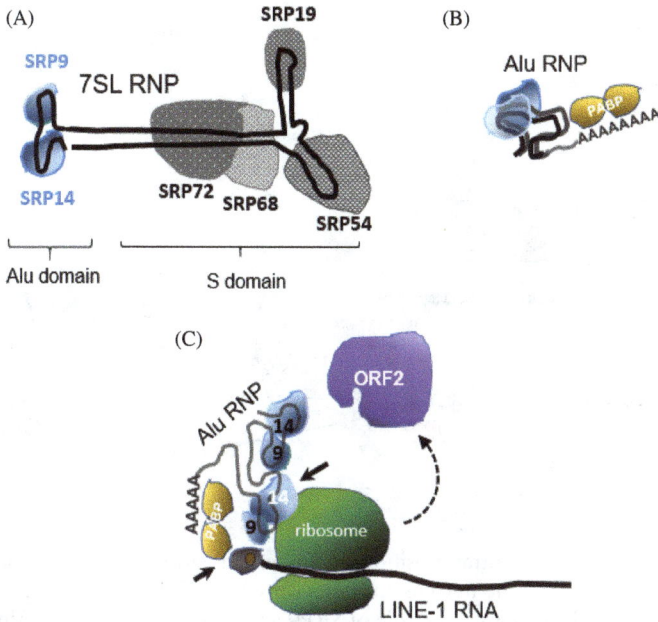

Figure 5.   The potential role of the 7SL SRP proteins and PABP in Alu retrotransposition. (A) Diagram of the 7SL RNP showing the SRP9 and SRP14 (blue) binding the "Alu domain" of the 7SL transcript. (B) The proposed model of the folded Alu RNA where both the left and right arms interact with the SRP9 and SRP14 proteins (blue) and PolyA-binding protein (PABP, yellow) bound to the 3′A-tail. (C) Proposed roles of the proteins in favoring the interaction of Alu with the L1 ORF2p. The SRP14 tethers the Alu RNP to the ribosome (arrow) bringing also the PABP in proximity to interact with the L1 RNA. This potentially allows the Alu to effectively parasitize the factors needed for retrotransposition.

with the Alu's poly-A tail and serves as a priming site for ORF2 to undergo reverse transcription. Previous studies have demonstrated that Alu's poly-A tail is critical for retrotransposition.[26,54,55]

Once inserted, not all Alu elements present in the genome are capable of generating inserts. In fact, only a few Alu loci appear to be the main contributors of the majority of the inserts in the human genome. This observation prompted the initial "Master" element or source element model.[39] What determines if a specific Alu element is active? Multiple factors influence the retrotransposition capability of any Alu locus.[25] One of the most important factors is the ability to generate the RNA template. In contrast to L1, Alu RNA lacks the

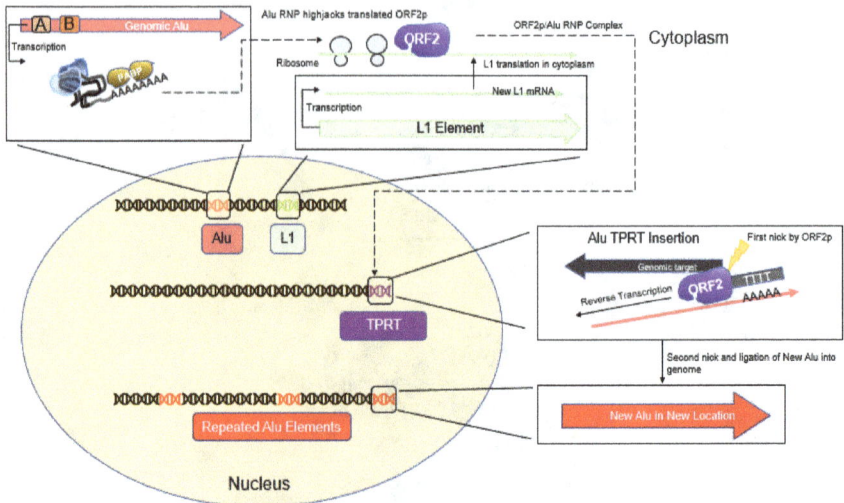

Figure 6.   The Alu retrotransposition life cycle. An Alu transcript is generated by RNA polymerase III in the nucleus and associates with SRP9/14 and PABP to form an RNP. The model proposes the role of SRP9 and SRP14 to target the Alu RNP to ribosomes increasing the probability of encountering an L1 transcript undergoing translation, which allows the "hijacking" of the ORF2p. In an unknown mechanism, the Alu returns to the nucleus to undergo insertion via TPRT. For TPRT to occur, the 3′A-tail of the Alu needs to base-pair with a T-rich region of the cleaved genomic DNA, which serves to prime reverse transcription. The site of insertion is relatively random as it targets AT-rich sites cleaved by the ORF2p endonuclease.

genetic material to ensure transcription. Although the Alu RNA sequence contains the internal Pol III A and B box, it lacks the upstream enhancer sequences.[56] Thus, a new insert depends on the genomic sequence at the site of insertion to provide the enhancer. Due to their relative random insertion, most Alu inserts will not land in a region containing an enhancer to support Pol III transcription. Even if an Alu transcript is generated, other internal sequences play a role by determining the ability to properly bind the SRP9/14 or have a sufficiently long A-tail to support TPRT.[54,55] Based on these components, individual Alu loci can be evaluated for estimated retrotransposition potential (ERP).[34]

Due to the lack of selection to maintain activity, Alu sequences mutate through time, becoming inactive. Usually, the first to degrade is the 3′-tail, followed by internal mutations in CpGs as well as other random mutations within the Alu body. Thus, only the younger Alu inserts (young subfamilies) still retain active copies.

2.3 *Alu expression*

RNA polymerase III drives transcription of Alu elements. Thus, Alu transcripts are not normally capped, spliced, or polyadenylated. The presence of a 3′ "A-tail" is a requirement for efficient retrotransposition of Alu elements.[54] Because RNA Pol III transcripts are not polyadenylated like typical Pol II mRNAs, Alu elements depend on their genomic 3′ A-tail sequences remaining intact to retain activity. As mentioned earlier, Alu transcripts can significantly vary in length due to the lack of an RNA Pol III terminator. Thus, the range of size variation of Alu transcripts is dictated by the specific loci being transcribed.

Determining Alu expression levels has proven to be a challenge. The biggest problem results from the high number of mRNAs that contain Alu sequences.[57,58] This contributes to a high background that obscures results. Some early estimates suggested that about 4% of human mRNAs harbor Alu sequences.[59,60] This background becomes exceptionally challenging, as common approaches to evaluate expression are limited in their ability to distinguish between RNA Pol III Alu transcripts and other transcripts that contain Alu sequences. For example, a standard RT-PCR approach using oligo dT for reverse transcription would generate the same product for both types of transcripts (Fig. 7(A)). Standard pipelines used in the analyzes of transcriptomic data from short read sequencing approaches also struggle to identify the Alu loci generating the transcript (Fig. 7(B)). Furthermore, reads containing flanking sequences are more likely derived from mRNAs containing an Alu sequence and not a true RNA Pol III Alu transcript (Fig 7(B)). Some published pipelines try to address these issues.[61,62] However, the biggest limitation of these pipelines is that they likely represent a significant underestimation, as most of the young potentially active Alu loci would not be captured. Due to the significant sequence homology, sequence reads from young Alu elements will perfectly align to multiple loci in the human

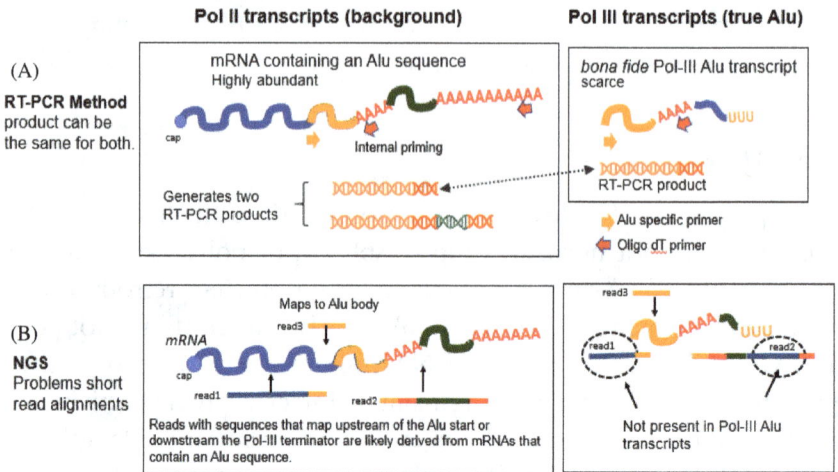

Figure 7. Limitations of PCR and short sequencing read methods in the ability to detect bona fide Alu pol-III generated transcripts. (A) A standard **RT-PCR** approach may yield a product that is derived from an mRNA containing an Alu sequence or it may be derived from a transcript of the Pol-III promoter of the Alu element. Because the Alu insert contains a 3′ A-stretch, it will provide an internal priming site for the oligo-dT. This will generate a product that is not distinguishable from one generated from a bona fide Pol-II transcript. (B) **Next-Generation Sequencing (NGS): Short sequence read transcriptomics** is complicated and limited in its ability to evaluate Alu transcripts. The main reason is that read mapping to the young active Alu elements (e.g., read 3 shown in yellow) will likely multimap, making it difficult to determine the specific locus that generated the RNA. In addition, read mapping to the Alu "body" (read 3), even if they mapped uniquely, would not distinguish between the RNA Pol III transcript and the mRNAs containing an Alu sequence. In addition, most of the unique mapping sequences from the flank (blue and green regions of the mRNA), if present in the reads (read 1 and read 2), are likely derived from an Alu-containing mRNA, as those sequences would not be present in a *bona fide* Pol III Alu transcript (shown on the right).

genome. If read multimapping is not allowed (settings in the software), these reads will be discarded and missing in the final data, and hence loss of information on young Alu elements. However, if multimapping is allowed, the reads will be randomly assigned to different loci, introducing then uncertainty as some silent Alu loci may be shown as expressing RNA. In general, most of the pipelines published select to discard reads that map to more than one locus.

The abundance of background becomes a concern in trusting the interpretation of published data that do not specifically address this issue. This abundant background also affects Northern blot analyzes reviewed in.[26] Using probes that hybridize to the Alu body will generate multiple bands of all sizes detecting both the RNA Pol III Alu transcripts that vary in size and any other transcript containing an Alu sequence. In addition, this probe also hybridizes with the very abundant 7SL RNA, which gives a strong uniform band.[63] For example, some publications analyzing Alu expression show northern blot outcomes with one strong band, reminiscent of detection of 7SL.[64] Another example of the pervasive background problem is a study by Hung *et al.* that led to the proposal of a role of retroelements in the development of systemic lupus erythematosus (SLE).[65] The authors report that Ro60 (a protein of unknown function associated with SLE that interacts with the non-coding Y RNAs) interacted with transcripts containing Alu sequences. However, further studies are needed to confirm the biological role of these transcripts. Other published data reporting Alu expression in tissues derive from *in situ* hybridization strategies, but provide no further evaluation to distinguish between 7SL and RNAs containing Alu sequences.[66,67] Recent publications have addressed some of these concerns by using strategies to separate the background from the RNA polymeras-III-transcribed SINEs.[68]

Overall, the amount of reliable data on Alu expression is limited. Several publications indicate low expression of Alu in human cell lines.[34,69] At the time of publication of this chapter, Alu expression in different human tissue remains mostly unknown.

3. Alu Elements and Human Disease

The impact of Alu elements on the human genome can be mostly attributed to two basic mechanisms: (1) genomic damage due to Alu retrotransposition activity, i.e., through insertional mutagenesis, and (2) post-insertional effects on cellular homeostasis and genetic instability (Fig. 8).

3.1 Alu insertional mutagenesis

The Human Genetic Mutation Database shows that *Alu* element insertions contribute to approximately 0.1% of human genetic

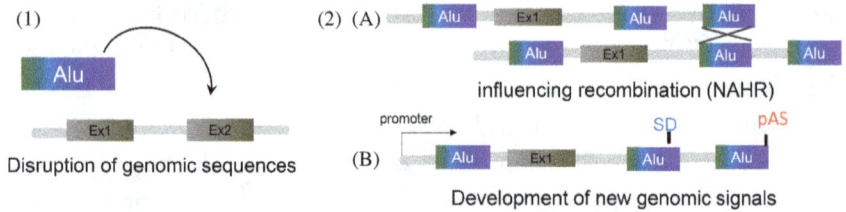

Figure 8. The two basic mechanisms of Alu-induced disease: (1) Alu insertional mutagenesis caused by the random insertion of Alu in the genome. (2) Post-insertional effects: the high abundance of Alu elements in the genome provides alternative sites for interaction during recombination events that can lead to (A) non-allelic homologous recombination (NAHR) or (B) by providing sequences that can alter function, e.g., introduction of new polyadenylation sites (pAS, shown in red) or splicing signals (SD, shown in blue).

diseases.[70] An estimated 0.3% of all human mutations are caused by germline *de novo* retrotransposition insertions. Among the three non-LTR retrotransposons, Alu has the highest retrotransposition rate (1 in every 20 births).[33] Due to the nature of its essentially random insertion,[71] Alu activity creates various impacts depending on the location of insertion. Some examples include an Alu insertion in the PKLR gene that causes pyruvate kinase deficiency,[72] insertion in the coding region of CLCN5 causing Dent's disease,[73] and an insertion in RP1 causing retinitis pigmentosa.[74] While a majority of the recent Alu insertions remain neutral, deleterious insertions occur through the disruption of coding sequence or regulatory sequences such as an insertion in an intron that affects splice sites. Furthermore, Alu insertions at 5′ and 3′ regions of a gene can negatively impact the gene expression process,[57,75] resulting in germline disease or increased susceptibility to cancer.[76–79] For example, an Alu insertion in exon 6 of MLH1 increases predisposition to Lynch Syndrome.[80] Diseases caused by *de novo* insertion of Alus in the germline range from blood disorders to cancer with over 60 catalogued cases in 2013[81] and counting and likely now over 100 cases. Reports of recurrent insertions in the same location may suggest variability in the susceptibility of the target gene. For example, two separate reports show Alu disruption of the BRCA2 gene.[77,78] Transposon density is suggested as a risk factor, as the ORF2p's canonical cleavage site (TTAAAA) is retained after the insertion of an Alu. Thus, the cleavage site can be reused creating tandem insertions of Alu elements.[82] In addition, the A-rich "tail" serves as a favorable sequence for the creation of new cleavage sites for new insertions. However, in some cases, the identification of recurring new insertions in the same gene could just be due to ascertainment bias.

Sometimes, the deletion of host DNA is associated with Alu insertions, a phenomenon termed Alu retrotransposition-mediated deletion (ARD).[83] A number of diseases have been linked to insertional-mediated deletion, for example, an ARD in the APC gene is associated with colon cancer.[84] Another report suggests that during primate evolution, ARD may have contributed to over 3000 deletions and loss of ~900 kb of DNA.[83] However, it is unclear if ARD occurs in a one-step mechanism or through an initial insertion followed by a secondary deletion step. Jahic *et al.*, argue in favor of the two-step mechanism supported by their careful study of two deletions in *SPAST* which are associated with hereditary spastic paraplegia type SPG4 (OMIM 604277).[85]

The vast majority of the reported Alu insertions associated with disease occurred in the germline. It is unclear if Alu retrotransposition occurs somatically. Although tissue culture studies demonstrate that Alu can jump in various cell types,[43] data on Alu somatic activity in human tissues are scarce. A few studies in cancers report Alu insertions associated with the disease.[86] The most compelling data derive from a systematic study of whole-genome sequencing data from tumors and matched normal samples that identified 10 high-confidence Alu somatic insertions.[87] However, further studies are needed to define the role of Alu activity in somatic tissues.

3.2 *Post-insertional impact of Alu insertion*

Once inserted into the genome, the mere presence of Alu elements can influence its genomic environment, potentially affecting function. One study demonstrated that polymorphic Alu elements are associated with risks for many diseases.[88] However, one study reports a protective role of a polymorphic Alu in the ACE gene and age-related macular degeneration[89] (AMD). Thus, Alu inserts can alter function in a variety of ways.

3.2.1 *Alu expression and disease*

Various studies implicate Alu expression in deregulation of cellular functions that lead to disease states.[90] In this review, the authors do not distinguish between Pol-III Alu transcripts and mRNAs containing Alu sequences. For example, multiple reports claim the accumulation of *Alu* RNA as a pathological causative factor of AMD.[66,91-93] Another published example reports an increased level of Alu transcripts in hepatocellular carcinoma tissue using *in situ* hybridization with a probe complementary to the AluYa5 body.[94] Alternatively, other studies have focused on the use of *in vitro* analyzes to study Alu RNA effects on translation.[95] In one example, the authors investigated the role of Alu RNPs and inhibition of translation initiation.[96] However, as mentioned in the previous section, measuring "true" RNA pol-III-driven Alu RNA expression is highly complex. Because of this problem, some of the published data on Alu expression correlated with disease may need careful re-evaluation.

3.2.2 *Post-insertional impact (non-allelic homologous recombination)*

While Alu insertional mutagenesis occasionally has deleterious effects through the deletion of host DNA and disruption of the regulatory DNA sequence, post-insertional recombination poses a greater threat to human genomic stability due to the larger size of the sequence alteration. An entire functional region may get deleted or duplicated. Due to their abundance and spread through all chromosomes (about one insertion every 3 kb, on average), Alu elements can contribute to a variety of misalignments during recombination with various results. Figure 9 illustrates some of the potential outcomes.

NAHR between Alu elements can mediate deletion of segments from as small as 300 bases up to tens of kilobases, as well as duplication, inversion, and translocation. This type of secondary genomic

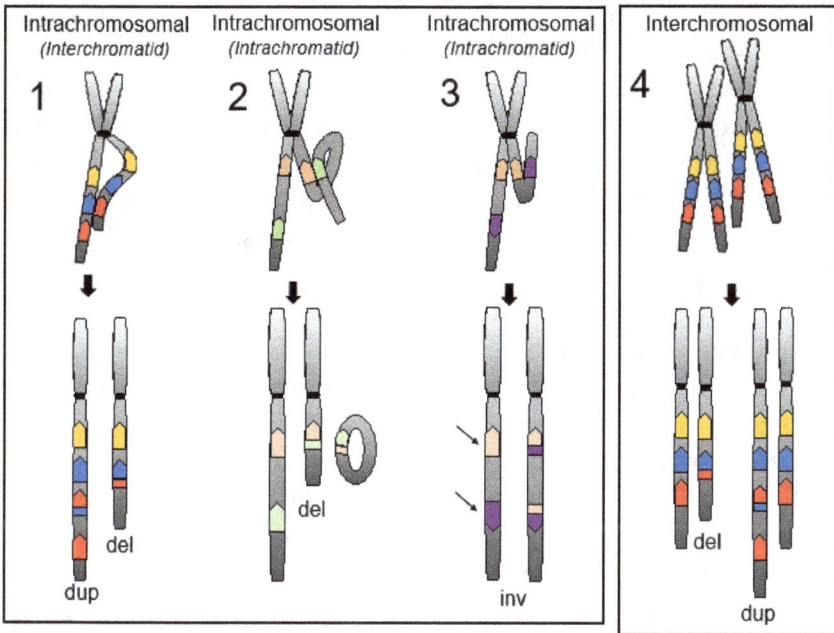

Figure 9. Schematic of Alu/Alu recombination models. The colored arrows represent Alu elements of a given subfamily. Alu/Alu recombination creates a chimeric Alu element (indicated by two different colors within the element). The breakage point can be anywhere within the Alu element. Definition, del: deletion, dup: duplication, and inv: inversion. (1) Intrachromosomal (Interchromatid) Alu/Alu recombination between two sister chromatids creates reciprocal deletion and duplication. (2) Intrachromosomal (Intrachromatid) Alu/Alu recombination produces only a deletion. (3) Intrachromosomal (Intrachromatid) Alu/Alu recombination between inverted Alu elements results in inversions. (4) Interchromosomal Alu/Alu recombination results in reciprocal deletion and duplication. If two non-homologous chromosomes are involved, a translocation can occur.

instability has a much bigger impact on gene disruption than the initial insertion of Alu elements. Comparative studies show that Alu-mediated deletions significantly affect genomes.[97,98] In addition, there are multiple examples of diseases frequently caused by Alu-mediated recombination.[99,100] Despite their relatively short size and sequence divergence, non-allelic Alu-mediated recombination events contribute to recombination hotspots[101] promoting genomic instability, including extensive DNA rearrangements associated with human disease.[86] Furthermore, detailed studies show that Alu/Alu-mediated rearrangements (AAMRs) are common, contributing to recombination hot spots.[102] Thus, some genes show susceptibility to independent recurrent AAMR-disease-causing events, e.g., those genes associated with the LDLR, AML, Fanconi anemia, and von Hippel-Lindau syndrome. This has led to the creation of tools such as the AluAluCNVpredictor, which allows for the assessment of AAMR.[103]

Alu-mediated NAHR events have been identified in a wide variety of genetic disorders, with Alu-mediated NAHR deletion being the predominant chromosomal variation.[104] While the lack of NAHR-mediated duplication events can be attributed to detection bias, genomic comparisons between humans and chimpanzees support the view that, overall, deletion occurs more often than duplication.[98,105] Multiple studies associate Alu-mediated NAHR with cancer predisposition, and evidence indicates that some genes are more prone to these recombination events than others. For example, recurrent Alu-mediated NAHRs have been observed in BRCA1, VHL, MLL1, MLH1, and MSH2 genes.[99,106,107] One specific example correlates the recurrent Alu-mediated loss of exon 7 of MSH2 with a high prevalence of early onset of colorectal, endometrial, and prostate cancers.[107] Models suggest that high Alu density contributes to the higher NAHR rate. High Alu density directly correlates with the presence of deletions in the genes associated with chronic myelogenous leukemia, acute lymphoblastic leukemia, and acute myelogenous leukemia.[108] Another example that supports this hypothesis comes from a study of MLH1 and MSH2 gene rearrangements.[109] Both genes have higher than average Alu density in their intronic sequence (21% for MLH1 and 34.2% for MSH2 compared to the 10% genomic average[109]). Li *et al.* also showed that only around a fourth of the recombination events from MLH1 contain hallmarks for unequal Alu recombination in comparison to 3/4 of the recombination events observed in MSH2 being Alu-mediated NAHRs.[109] Studies on mutations of the BRCA1 gene (41.5% Alu density[110]) reported that about 70% of deletion were Alu-mediated NAHR.[106] Another contributing factor for Alu-mediated NAHR is the shared homology between Alu sequences.[111] Studies on genomic data support this observation, as Alu inserts that retain sequences closest to the consensus (such as the youngest Alu subfamily, AluY) show a disproportional recombination activity relative to their density.[98] The resulting higher homology allows for an increased recombination rate between the younger families.

3.2.3 *Post-insertional impact (epigenetics, transcription regulation)*

Once inserted into the genome, Alu sequences can influence promoters of genes by introducing new signals or contribute to epigenetic changes (Fig. 10). These sequences can be already present in the Alu or they can evolve through time. Note that Alu elements in inverted orientations (i.e., the complementary strand) can also influence genes in this manner. One such example is the adaptation of an Alu insert in the BRCA-1 gene that diverged and acquired the ability to function as an estrogen receptor-dependent enhancer.[112]

Alu elements can regulate RNA transcription through DNA methylation due to the abundance of CpG residues present in its sequence.[113] Alu elements are probably a significant source of CpG islands in the genome.[114] Estimates indicate that 25% of all DNA methylation sites locate to Alu sequences.[115] One report illustrates this type of effect by demonstrating the regulation of the MIEN1 gene by the methylation of an Alu in its promoter.[116] Another example reports

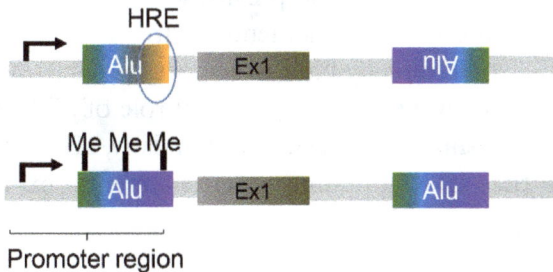

Figure 10. Schematic of the influence of Alu elements in or near promoter regions influencing transcription of genes. Top: Random mutational changes in Alu sequences have sometimes led to the creation of novel Hormone Receptor Elements (HRE, shown in yellow) and changed the transcription regulation of genes.[112] Bottom: Alu elements being rich in CpGs provide potential regions for methylation and regulation of genes when present in or near promoter regions. (Me: methylation).

the epigenetic influence of an Alu element variant in the POMC gene, associated with childhood obesity.[117] Some of the Alu-mediated methylation changes correlate with cancerous states, as highlighted in a study showing the spread of Alu methylation to the promoter of the MLH1 gene in gastrointestinal cancer.[118]

Furthermore, Alu sequences can contain several transcription factor (TF) binding sites, some of which are proposed to play a role in regulation and development.[119] While some of these sites are shared between several Alu subfamilies, other binding sites originate from the accumulation of sequence changes occurring post insertion. As previously mentioned, an Alu element in BRCA-1 diverged to function as an estrogen receptor-dependent enhancer.[112] This work provided strong evidence that a subset of Alu elements can confer estrogen responsiveness upon a nearby promoter. In addition, an *in silico* analysis of progesterone response elements (PRE) in progesterone receptor up-regulated promoters showed that Alu repeats can serve as co-response elements.[120] Interestingly, bioinformatic studies found that p53 binding sites and p53 response elements resided in Alu repeats.[121] The authors propose that the presence of p53 RE in intronic Alu repeats may regulate Alu transcription and possibly also influence host gene expression. However, to determine the exact role of these p53 binding sites present within Alu sequences, further studies are needed. Earlier studies had previously shown that p53 binding sites can also be found in L1 elements.[122] In particular, the authors report that p53 could directly bind to a short 15-nucleotide sequence in the L1 promoter and suggest a potential role of p53 in protecting against genomic insults by this element. Similarly, p53 sites have been reported in the LTR-containing retroelements in the human genome.[123]

3.2.4   *Post-insertional impact (RNA processing, exonization, and polyadenylation)*

Apart from Alu's influence on RNA transcription, it also interferes with RNA processing activities such as splicing and polyadenylation leading to altered expression.[124] A large number of genes contain Alu elements in introns[125] and in their 3' UTRs.[126] Figure 11 illustrates

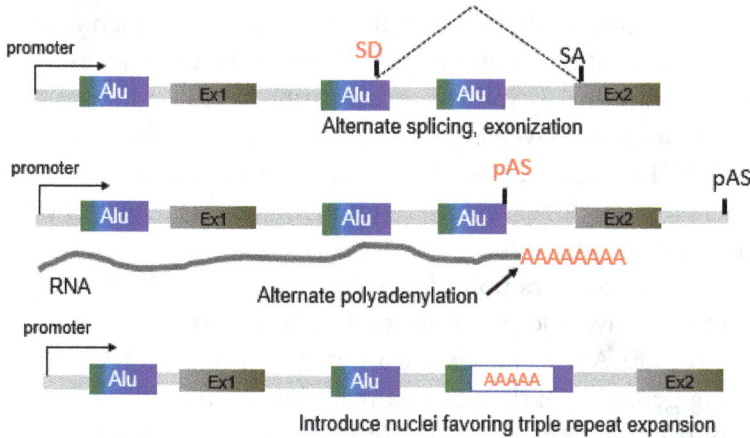

Figure 11. Alu insertions affecting RNA processing or stability. Alu insertions introduce sequences that may contain splicing signals (splice donors (SD) or splice accetors (SA)), polyadenylation signals (pAS), or provide A-rich stretches that serve as nuclei for expansion.

several of the mechanisms by which Alu sequences can affect RNA processing or genetic instability. Many of the Alu sequences have been shown to provide alternate splice sites[127] and alternate polyadenylation sites.[128] Most Alu elements contain motifs highly similar to the splicing signals. Thus, cryptic splice sites can be easily created by a point mutation in the Alu element and contribute to alternate splicing. For example, an Alu element can harbor up to 10 potential 5′ splice sites and 13 potential 3′ splice sites.[115] Published data showed examples of common inherited polymorphic Alu insertions that altered splicing of five genes.[129] Similarly, an Alu insert in the third intron of the ornithine aminotransferase gene contributes to a cryptic splice site, that when used, causes the disruption of gene function, resulting in eye disease.[130] In another example, an intronic AluYb9 in the Factor VIII gene causes exon skipping leading to hemophilia A.[131] More recently, a polymorphic Alu insert in CD58 associated with multiple sclerosis was shown to alter splicing of the mRNA by skipping exon 3.[129]

Retention of intronic sequences due to Alu is referred to as exonization. Older Alu subfamilies are usually more involved in the exonization process due to the sufficient accumulation of mutations to create the cryptic splice site.[132] Moreover, the complementary strand of an Alu element contains a 5′ poly-T stretch that can act as a pyrimidine-rich tract for 3′ splice site formation.[133] Alu exonization can create a new exon, as well as add a new coding sequence to an existing exon.[134–136] In certain cases, this process can introduce a stop codon within a protein-coding region, resulting in the shortening of protein, which can be deleterious for the organism.

Alu also introduces polyadenylation signals, which are less prevalent yet still have the potential to interfere with RNA processing. About 10,000 Alu elements are located in the 3′ UTR of protein encoding genes.[137] Of these, a vast majority (99%) of the polyadenylation contributing Alu elements are in the forward orientation. One example of an *Alu* polyadenylation site affecting gene function is an Alu insertion present in exon 7 of the calcium-sensing receptor, CaSR. Polyadenylation at this Alu causes premature termination as it provides a stop codon signal leading to hypocalciuric hypercalcemia and neonatal severe hyperparathyroidism.[138–140] In addition, these A-rich stretches can also favor expansion into microsatellites[141] or triplet repeats.[142,143]

3.2.5 *Post-insertional impact (RNA editing)*

The presence of Alu elements within an RNA can also influence how RNA editing enzymes process the transcripts. RNA editing is a process by which cells can make changes to certain nucleotides in the genome and generate protein diversity. Adenosine-to-Inosine RNA modification is the most common type of double-strand RNA editing that occurs in mammals, and the family of ADAR enzymes (ADAR1 and ADAR2) catalyzes this process. The amount of editing sites in the substrate depends on the length of the RNA duplex. Shorter sequences with interruptions, like loop structures, are only edited in one site, while longer substrates can have all of their adenosines edited.[144]

In transcripts, long duplex structures can form when an inverted Alu is in close proximity to another Alu repeat (see Fig. 12). About 700,000 Alu repeats out of the 1.1 million fit into this category and

RNA with inverted Alu elements

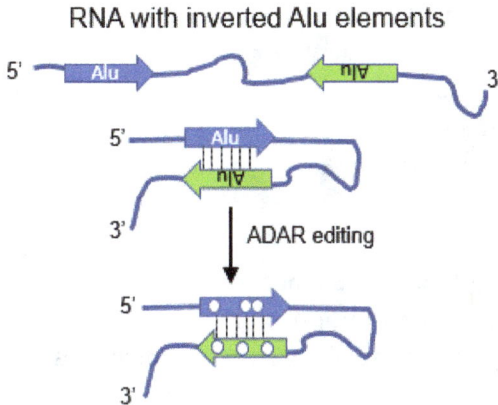

Figure 12. Inverted Alu elements in mRNAs contribute to the formation of double-strand RNA structures that are targets of editing processes. Editing of the mRNA by ADAR may activate alternative splicing, nuclear retention, transcript stability, or interfere with translation.

are subject to A-to-I editing.[115] The consequence of this A-to-I editing ranges from creating cryptic splice sites to disrupting functional sequence within the RNA.

A comprehensive study determined that the RNA editing of Alu sequences likely occurs globally.[145,146] Multiple reports show associations between Alu-mediated RNA editing and neurological diseases[147] as well as rheumatoid arthritis.[148] Editing also may play a role in cancer, where Alu-dependent editing and transcriptional activity of GLI1 promote malignant regeneration of multiple myeloma.[149] Furthermore, editing not only occurs in protein-coding regions of mRNAs but also occurs frequently in non-coding regions containing inverted Alu repeats. For example, some edited microRNA (miRNA) can lead to either the reduction of expression or the altered function of mature miRNAs.[150]

Long duplex RNA is a potent stressor on the cell and can trigger an immune response. In the case of viral RNA, this is resolved through the autophosphorylation of protein kinase R (PKR), which in turn can cause translational arrest and cell apoptosis. A similar interaction between PKR and Alu dsRNA was also documented during the early phase of mitosis.[151] The mechanism of how Alu dsRNA prevents activation of cytosolic PKR remains unclear, but a recent report has identified an essential role played by ADAR1 in the process.[152] Further studies are needed to determine the full impact of Alu editing and human disease.

4. Summary

With over one million Alu elements in the human genome, it is clear that Alu elements play an important role by contributing to diversity and to a variety of genetic diseases. Alu contributes to human disease both actively via gene disruptions through insertional mutation and once inserted by passively influencing its genomic surroundings. The advent of new technologies that can address both whole genomes and individual cell variation (somatic polymorphisms) will provide more insight into the frequency and impact of Alu elements on human disease. However, caution is needed in the interpretation of data, as the abundance of Alu sequences within genes creates multiple sources for background issues that can obscure results.

References

1. Lander ES, Linton LM, Birren B, *et al*. Initial sequencing and analysis of the human genome. *Nature* 2001;409:860–921.
2. de Koning AP, Gu W, Castoe TA, Batzer MA, Pollock DD. Repetitive elements may comprise over two-thirds of the human genome. *PLoS Genet* 2011;7:e1002384.
3. Finnegan DJ. Eukaryotic transposable elements and genome evolution. *Trends Genet* 1989;5:103–7.
4. Feschotte C, Pritham EJ. DNA transposons and the evolution of eukaryotic genomes. *Annu Rev Genet* 2007;41:331–68.
5. Pace JK, 2nd, Feschotte C. The evolutionary history of human DNA transposons: evidence for intense activity in the primate lineage. *Genome Res* 2007;17:422–32.
6. Platt RN, 2nd, Vandewege MW, Ray DA. Mammalian transposable elements and their impacts on genome evolution. *Chromosome Res* 2018;26:25–43.

7. Weiner AM, Deininger PL, Efstratiadis A. Nonviral retroposons: genes, pseudogenes, and transposable elements generated by the reverse flow of genetic information. *Annu Rev Biochem* 1986;55: 631–61.
8. Schmitz J, Churakov G, Zischler H, Brosius J. A novel class of mammalian-specific tailless retropseudogenes. *Genome Res* 2004;14:1911–5.
9. Boeke JD, Garfinkel DJ, Styles CA, Fink GR. Ty elements transpose through an RNA intermediate. *Cell* 1985;40:491–500.
10. Havecker ER, Gao X, Voytas DF. The diversity of LTR retrotransposons. *Genome Biol* 2004;5:225.
11. Ke N, Gao X, Keeney JB, Boeke JD, Voytas DF. The yeast retrotransposon Ty5 uses the anticodon stem-loop of the initiator methionine tRNA as a primer for reverse transcription. *RNA* 1999;5:929–38.
12. Esnault C, Maestre J, Heidmann T. Human LINE retrotransposons generate processed pseudogenes. *Nat Genet* 2000;24:363–7.
13. Swergold GD. Identification, characterization, and cell specificity of a human LINE-1 promoter. *Mol Cell Biol* 1990;10:6718–29.
14. Speek M. Antisense promoter of human L1 retrotransposon drives transcription of adjacent cellular genes. *Mol Cell Biol* 2001;21: 1973–85.
15. Denli AM, Narvaiza I, Kerman BE, *et al.* Primate-specific ORF0 contributes to retrotransposon-mediated diversity. *Cell* 2015;163:583–93.
16. Hohjoh H, Singer MF. Sequence-specific single-strand RNA binding protein encoded by the human LINE-1 retrotransposon. *EMBO J* 1997;16:6034–43.
17. Feng Q, Moran JV, Kazazian HH, Jr., Boeke JD. Human L1 retrotransposon encodes a conserved endonuclease required for retrotransposition. *Cell* 1996;87:905–16.
18. Mathias SL, Scott AF, Kazazian HH, Jr., Boeke JD, Gabriel A. Reverse transcriptase encoded by a human transposable element. *Science* 1991;254:1808–10.
19. Brouha B, Schustak J, Badge RM, *et al.* Hot L1s account for the bulk of retrotransposition in the human population. *Proc Natl Acad Sci USA* 2003;100:5280–5.
20. Streva VA, Jordan VE, Linker S, Hedges DJ, Batzer MA, Deininger PL. Sequencing, identification and mapping of primed L1 elements (SIMPLE) reveals significant variation in full length L1 elements between individuals. *BMC Genomics* 2015;16:220.

21. Sun FJ, Fleurdepine S, Bousquet-Antonelli C, Caetano-Anolles G, Deragon JM. Common evolutionary trends for SINE RNA structures. *Trends Genet* 2007;23:26–33.

22. Kramerov DA, Vassetzky NS. Short retroposons in eukaryotic genomes. *Int Rev Cytol* 2005;247:165–221.

23. Walter P, Blobel G. Signal recognition particle contains a 7S RNA essential for protein translocation across the endoplasmic reticulum. *Nature* 1982;299:691–8.

24. Ullu E, Tschudi C. Alu sequences are processed 7SL RNA genes. *Nature* 1984;312:171–2.

25. Comeaux MS, Roy-Engel AM, Hedges DJ, Deininger PL. Diverse cis factors controlling Alu retrotransposition: what causes Alu elements to die? *Genome Res* 2009;19:545–55.

26. Roy-Engel AM. LINEs, SINEs and other retroelements: do birds of a feather flock together? Front Biosci (Landmark Ed) 2012;17: 1345–61.

27. Kriegs JO, Churakov G, Jurka J, Brosius J, Schmitz J. Evolutionary history of 7SL RNA-derived SINEs in Supraprimates. *Trends Genet* 2007;23:158–61.

28. Quentin Y. Fusion of a free left Alu monomer and a free right Alu monomer at the origin of the Alu family in the primate genomes. *Nucleic Acids Res* 1992;20:487–93.

29. Zietkiewicz E, Richer C, Makalowski W, Jurka J, Labuda D. A young Alu subfamily amplified independently in human and African great apes lineages. *Nucleic Acids Res* 1994;22:5608–12.

30. Quentin Y. Origin of the Alu family: a family of Alu-like monomers gave birth to the left and the right arms of the Alu elements. *Nucleic Acids Res* 1992;20:3397–401.

31. Kojima KK. Alu monomer revisited: recent generation of Alu monomers. *Mol Biol Evol* 2011;28:13–5.

32. Deininger PL, Batzer MA, Hutchison CA, 3rd, Edgell MH. Master genes in mammalian repetitive DNA amplification. *Trends Genet* 1992;8:307–11.

33. Cordaux R, Hedges DJ, Herke SW, Batzer MA. Estimating the retrotransposition rate of human Alu elements. *Gene* 2006;373:134–7.

34. Oler AJ, Traina-Dorge S, Derbes RS, Canella D, Cairns BR, Roy-Engel AM. Alu expression in human cell lines and their retrotranspositional potential. *Mob DNA* 2012;3:11.

35. Sveinbjornsson JI, Halldorsson BV. PAIR: polymorphic Alu insertion recognition. *BMC Bioinformatics* 2012;13 Suppl 6:S7.
36. Batzer MA, Deininger PL. Alu repeats and human genomic diversity. *Nat Rev Genet* 2002;3:370–9.
37. Hedges DJ, Callinan PA, Cordaux R, Xing J, Barnes E, Batzer MA. Differential alu mobilization and polymorphism among the human and chimpanzee lineages. *Genome Res* 2004;14:1068–75.
38. Ray DA, Walker JA, Batzer MA. Mobile element-based forensic genomics. *Mutat Res* 2007;616:24–33.
39. Shen MR, Batzer MA, Deininger PL. Evolution of the master Alu gene(s). *J Mol Evol* 1991;33:311–20.
40. Han K, Xing J, Wang H, *et al.* Under the genomic radar: the stealth model of Alu amplification. *Genome Res* 2005;15:655–64.
41. Watkins WS, Feusier JE, Thomas J, Goubert C, Mallick S, Jorde LB. The Simons Genome Diversity Project: A global analysis of mobile element diversity. *Genome Biol Evol* 2020;12:779–94.
42. Dewannieux M, Esnault C, Heidmann T. LINE-mediated retrotransposition of marked Alu sequences. *Nat Genet* 2003;35:41–8.
43. Wallace N, Wagstaff BJ, Deininger PL, Roy-Engel AM. LINE-1 ORF1 protein enhances Alu SINE retrotransposition. *Gene* 2008;419:1–6.
44. Thomas Y, Bui N, Strub K. A truncation in the 14 kDa protein of the signal recognition particle leads to tertiary structure changes in the RNA and abolishes the elongation arrest activity of the particle. *Nucleic Acids Res* 1997;25:1920–9.
45. Terzi L, Pool MR, Dobberstein B, Strub K. Signal recognition particle Alu domain occupies a defined site at the ribosomal subunit interface upon signal sequence recognition. *Biochemistry* 2004;43:107–17.
46. Ahl V, Keller H, Schmidt S, Weichenrieder O. Retrotransposition and crystal structure of an Alu RNP in the ribosome-stalling conformation. *Mol Cell* 2015;60:715–27.
47. Bennett EA, Keller H, Mills RE, *et al.* Active Alu retrotransposons in the human genome. *Genome Res* 2008;18:1875–83.
48. Sarrowa J, Chang DY, Maraia RJ. The decline in human Alu retroposition was accompanied by an asymmetric decrease in SRP9/14 binding to dimeric Alu RNA and increased expression of small cytoplasmic Alu RNA. *Mol Cell Biol* 1997;17:1144–51.
49. Boeke JD. LINEs and Alus--the polyA connection. *Nat Genet* 1997; 16:6–7.

50. West N, Roy-Engel AM, Imataka H, Sonenberg N, Deininger P. Shared protein components of SINE RNPs. *J Mol Biol* 2002;321: 423–32.

51. Gingras AC, Raught B, Sonenberg N. eIF4 initiation factors: effectors of mRNA recruitment to ribosomes and regulators of translation. *Annu Rev Biochem* 1999;68:913–63.

52. Dai L, Taylor MS, O'Donnell KA, Boeke JD. Poly(A) binding protein C1 is essential for efficient L1 retrotransposition and affects L1 RNP formation. *Mol Cell Biol* 2012;32:4323–36.

53. Luan DD, Korman MH, Jakubczak JL, Eickbush TH. Reverse transcription of R2Bm RNA is primed by a nick at the chromosomal target site: a mechanism for non-LTR retrotransposition. *Cell* 1993;72: 595–605.

54. Dewannieux M, Heidmann T. Role of poly(A) tail length in Alu retrotransposition. *Genomics* 2005;86:378–81.

55. Roy-Engel AM, Salem AH, Oyeniran OO, *et al.* Active Alu element "A-tails": size does matter. *Genome Res* 2002;12:1333–44.

56. Roy AM, West NC, Rao A, *et al.* Upstream flanking sequences and transcription of SINEs. *J Mol Biol* 2000;302:17–25.

57. Landry JR, Medstrand P, Mager DL. Repetitive elements in the 5′ untranslated region of a human zinc-finger gene modulate transcription and translation efficiency. *Genomics* 2001;76:110–6.

58. Daskalova E, Baev V, Rusinov V, Minkov I. 3′UTR-located ALU elements: donors of potential miRNA target sites and mediators of network miRNA-based regulatory interactions. *Evol Bioinform Online* 2007;2:103–20.

59. Jasinska A, Krzyzosiak WJ. Repetitive sequences that shape the human transcriptome. *FEBS Lett* 2004;567:136–41.

60. Yulug IG, Yulug A, Fisher EM. The frequency and position of Alu repeats in cDNAs, as determined by database searching. *Genomics* 1995;27:544–8.

61. Carnevali D, Dieci G. Identification of RNA polymerase III-transcribed SINEs at single-locus resolution from RNA sequencing data. *Noncoding RNA* 2017;3.

62. Conti A, Carnevali D, Bollati V, Fustinoni S, Pellegrini M, Dieci G. Identification of RNA polymerase III-transcribed Alu loci by computational screening of RNA-Seq data. *Nucleic Acids Res* 2015;43:817–35.

63. Paulson KE, Schmid CW. Transcriptional inactivity of Alu repeats in HeLa cells. *Nucleic Acids Res* 1986;14:6145–58.

64. Mariner PD, Walters RD, Espinoza CA, *et al.* Human Alu RNA is a modular transacting repressor of mRNA transcription during heat shock. *Mol Cell* 2008;29:499–509.

65. Hung T, Pratt GA, Sundararaman B, *et al.* The Ro60 autoantigen binds endogenous retroelements and regulates inflammatory gene expression. *Science* 2015;350:455–9.

66. Kaneko H, Dridi S, Tarallo V, *et al.* DICER1 deficit induces Alu RNA toxicity in age-related macular degeneration. *Nature* 2011;471:325–30.

67. Tarallo V, Hirano Y, Gelfand BD, *et al.* DICER1 loss and Alu RNA induce age-related macular degeneration via the NLRP3 inflammasome and MyD88. *Cell* 2012;149:847–59.

68. Mori Y, Ichiyanagi K. melRNA-seq for expression analysis of SINE RNAs and other medium-length non-coding RNAs. *Mob DNA* 2021; 12:15.

69. Shaikh TH, Roy AM, Kim J, Batzer MA, Deininger PL. cDNAs derived from primary and small cytoplasmic Alu (scAlu) transcripts. *J Mol Biol* 1997;271:222–34.

70. Deininger PL, Batzer MA. Alu repeats and human disease. *Mol Genet Metab* 1999;67:183–93.

71. Arcot SS, Shaikh TH, Kim J, *et al.* Sequence diversity and chromosomal distribution of "young" Alu repeats. *Gene* 1995;163:273–8.

72. Lesmana H, Dyer L, Li X, *et al.* Alu element insertion in PKLR gene as a novel cause of pyruvate kinase deficiency in Middle Eastern patients. *Hum Mutat* 2018;39:389–93.

73. Claverie-Martin F, Gonzalez-Acosta H, Flores C, Anton-Gamero M, Garcia-Nieto V. De novo insertion of an Alu sequence in the coding region of the CLCN5 gene results in Dent's disease. *Hum Genet* 2003;113:480–5.

74. Nishiguchi KM, Fujita K, Ikeda Y, *et al.* A founder Alu insertion in RP1 gene in Japanese patients with retinitis pigmentosa. *Jpn J Ophthalmol* 2020;64:346–50.

75. Sobczak K, Krzyzosiak WJ. Structural determinants of BRCA1 translational regulation. J Biol Chem 2002;277:17349–58.

76. Halling KC, Lazzaro CR, Honchel R, *et al.* Hereditary desmoid disease in a family with a germline Alu I repeat mutation of the APC gene. *Hum Hered* 1999;49:97–102.

77. Miki Y, Katagiri T, Kasumi F, Yoshimoto T, Nakamura Y. Mutation analysis in the BRCA2 gene in primary breast cancers. *Nat Genet* 1996;13:245–7.

78. Teugels E, De Brakeleer S, Goelen G, Lissens W, Sermijn E, De Greve J. De novo Alu element insertions targeted to a sequence common to the BRCA1 and BRCA2 genes. *Hum Mutat* 2005;26:284.
79. Wallace MR, Andersen LB, Saulino AM, Gregory PE, Glover TW, Collins FS. A de novo Alu insertion results in neurofibromatosis type 1. *Nature* 1991;353:864–6.
80. Solassol J, Larrieux M, Leclerc J, *et al.* Alu element insertion in the MLH1 exon 6 coding sequence as a mutation predisposing to Lynch syndrome. *Hum Muta*t 2019;40:716–20.
81. Kaer K, Speek M. Retroelements in human disease. *Gene* 2013;518:231–41.
82. El-Sawy M, Deininger P. Tandem insertions of Alu elements. *Cytogenet Genome Res* 2005;108:58–62.
83. Callinan PA, Wang J, Herke SW, Garber RK, Liang P, Batzer MA. Alu retrotransposition-mediated deletion. *J Mol Biol* 2005;348:791–800.
84. Tuohy TM, Done MW, Lewandowski MS, *et al.* Large intron 14 rearrangement in APC results in splice defect and attenuated FAP. *Hum Genet* 2010;127:359–69.
85. Jahic A, Erichsen AK, Deufel T, Tallaksen CM, Beetz C. A polymorphic Alu insertion that mediates distinct disease-associated deletions. *Eur J Hum Genet* 2016;24:1371–4.
86. Ade C, Roy-Engel AM, Deininger PL. Alu elements: an intrinsic source of human genome instability. *Curr Opin Virol* 2013;3:639–45.
87. Lee E, Iskow R, Yang L, *et al.* Landscape of somatic retrotransposition in human cancers. *Science* 2012;337:967–71.
88. Payer LM, Steranka JP, Yang WR, *et al.* Structural variants caused by Alu insertions are associated with risks for many human diseases. *Proc Natl Acad Sci U S A* 2017;114:E3984–92.
89. Hamdi HK, Reznik J, Castellon R, *et al.* Alu DNA polymorphism in ACE gene is protective for age-related macular degeneration. *Biochem Biophys Res Commun* 2002;295:668–72.
90. Zhang L, Chen JG, Zhao Q. Regulatory roles of Alu transcript on gene expression. *Exp Cell Res* 2015;338:113–8.
91. Kaarniranta K, Pawlowska E, Szczepanska J, Blasiak J. DICER1 in the pathogenesis of age-related macular degeneration (AMD) — Alu RNA accumulation versus miRNA dysregulation. *Aging Dis* 2020;11:851–62.
92. Kim Y, Tarallo V, Kerur N, *et al.* DICER1/Alu RNA dysmetabolism induces Caspase-8-mediated cell death in age-related macular degeneration. *Proc Natl Acad Sci USA* 2014;111:16082–7.

93. Yoshida H, Matsushita T, Kimura E, *et al.* Systemic expression of Alu RNA in patients with geographic atrophy secondary to age-related macular degeneration. *PLoS One* 2019;14:e0220887.

94. Tang RB, Wang HY, Lu HY, *et al.* Increased level of polymerase III transcribed Alu RNA in hepatocellular carcinoma tissue. *Mol Carcinog* 2005;42:93–6.

95. Chang DY, Newitt JA, Hsu K, Bernstein HD, Maraia RJ. A highly conserved nucleotide in the Alu domain of SRP RNA mediates translation arrest through high affinity binding to SRP9/14. *Nucleic Acids Res* 1997;25:1117–22.

96. Ivanova E, Berger A, Scherrer A, Alkalaeva E, Strub K. Alu RNA regulates the cellular pool of active ribosomes by targeted delivery of SRP9/14 to 40S subunits. *Nucleic Acids Res* 2015;43:2874–87.

97. Han K, Lee J, Meyer TJ, *et al.* Alu recombination-mediated structural deletions in the chimpanzee genome. *PLoS Genet* 2007;3:1939–49.

98. Sen SK, Han K, Wang J, *et al.* Human genomic deletions mediated by recombination between Alu elements. *Am J Hum Genet* 2006;79: 41–53.

99. Franke G, Bausch B, Hoffmann MM, *et al.* Alu-Alu recombination underlies the vast majority of large VHL germline deletions: molecular characterization and genotype-phenotype correlations in VHL patients. *Hum Mutat* 2009;30:776–86.

100. Gu S, Yuan B, Campbell IM, *et al.* Alu-mediated diverse and complex pathogenic copy-number variants within human chromosome 17 at p13.3. *Hum Mol Genet* 2015;24:4061–77.

101. McVean G. What drives recombination hotspots to repeat DNA in humans? *Philos Trans R Soc Lond B Biol Sci* 2010;365:1213–8.

102. Lupski JR. Hotspots of homologous recombination in the human genome: not all homologous sequences are equal. *Genome Biol* 2004;5:242.

103. Song X, Beck CR, Du R, *et al.* Predicting human genes susceptible to genomic instability associated with Alu/Alu-mediated rearrangements. *Genome Res* 2018;28:1228–42.

104. Belancio VP, Roy-Engel AM, Deininger PL. All y'all need to know 'bout retroelements in cancer. *Semin Cancer Biol* 2010;20:200–10.

105. Han K, Lee J, Meyer TJ, Remedios P, Goodwin L, Batzer MA. L1 recombination-associated deletions generate human genomic variation. *Proc Natl Acad Sci USA* 2008;105:19366–71.

106. del Valle J, Feliubadalo L, Lazaro C. Comments on: Sluiter MD and van Rensburg EJ, Large genomic rearrangements of the BRCA1 and BRCA2 genes: review of the literature and report of a novel BRCA1 mutation. *Breast Cancer Res Treat* 2010;124:295–6.

107. Perez-Cabornero L, Borras Flores E, Infante Sanz M, *et al.* Characterization of new founder Alu-mediated rearrangements in MSH2 gene associated with a Lynch syndrome phenotype. *Cancer Prev Res (Phila)* 2011;4:1546–55.

108. Kolomietz E, Meyn MS, Pandita A, Squire JA. The role of Alu repeat clusters as mediators of recurrent chromosomal aberrations in tumors. *Genes Chromosom Cancer* 2002;35:97–112.

109. Li L, McVety S, Younan R, *et al.* Distinct patterns of germ-line deletions in MLH1 and MSH2: the implication of Alu repetitive element in the genetic etiology of Lynch syndrome (HNPCC). *Hum Mutat* 2006;27:388.

110. Smith TM, Lee MK, Szabo CI, *et al.* Complete genomic sequence and analysis of 117 kb of human DNA containing the gene BRCA1. *Genome Res* 1996;6:1029–49.

111. Morales ME, White TB, Streva VA, DeFreece CB, Hedges DJ, Deininger PL. The contribution of alu elements to mutagenic DNA double-strand break repair. *PLoS Genet* 2015;11:e1005016.

112. Norris J, Fan D, Aleman C, *et al.* Identification of a new subclass of Alu DNA repeats which can function as estrogen receptor-dependent transcriptional enhancers. *J Biol Chem* 1995;270:22777–82.

113. Oei SL, Babich VS, Kazakov VI, Usmanova NM, Kropotov AV, Tomilin NV. Clusters of regulatory signals for RNA polymerase II transcription associated with Alu family repeats and CpG islands in human promoters. *Genomics* 2004;83:873–82.

114. Kang MI, Rhyu MG, Kim YH, *et al.* The length of CpG islands is associated with the distribution of Alu and L1 retroelements. *Genomics* 2006;87:580–90.

115. Daniel C, Behm M, Ohman M. The role of Alu elements in the cis-regulation of RNA processing. *Cell Mol Life Sci* 2015;72:4063–76.

116. Rajendiran S, Gibbs LD, Van Treuren T, Klinkebiel DL, Vishwanatha JK. MIEN1 is tightly regulated by SINE Alu methylation in its promoter. *Oncotarget* 2016;7:65307–19.

117. Kuehnen P, Mischke M, Wiegand S, *et al.* An Alu element-associated hypermethylation variant of the POMC gene is associated with childhood obesity. *PLoS Genet* 2012;8:e1002543.

118. Wang X, Fan J, Liu D, Fu S, Ingvarsson S, Chen H. Spreading of Alu methylation to the promoter of the MLH1 gene in gastrointestinal cancer. *PLoS One* 2011;6:e25913.
119. Polak P, Domany E. Alu elements contain many binding sites for transcription factors and may play a role in regulation of developmental processes. *BMC Genomics* 2006;7:133.
120. Jacobsen BM, Jambal P, Schittone SA, Horwitz KB. ALU repeats in promoters are position-dependent co-response elements (coRE) that enhance or repress transcription by dimeric and monomeric progesterone receptors. *Mol Endocrinol* 2009;23:989–1000.
121. Cui F, Sirotin MV, Zhurkin VB. Impact of Alu repeats on the evolution of human p53 binding sites. *Biol Direct* 2011;6:2.
122. Harris CR, Dewan A, Zupnick A, *et al.* p53 responsive elements in human retrotransposons. *Oncogene* 2009;28:3857–65.
123. Wang T, Zeng J, Lowe CB, *et al.* Species-specific endogenous retroviruses shape the transcriptional network of the human tumor suppressor protein p53. *Proc Natl Acad Sci USA* 2007;104:18613–8.
124. Sela N, Mersch B, Gal-Mark N, Lev-Maor G, Hotz-Wagenblatt A, Ast G. Comparative analysis of transposed element insertion within human and mouse genomes reveals Alu's unique role in shaping the human transcriptome. *Genome Biol* 2007;8:R127.
125. Tsirigos A, Rigoutsos I. Alu and b1 repeats have been selectively retained in the upstream and intronic regions of genes of specific functional classes. *PLoS Comput Biol* 2009;5:e1000610.
126. Grover D, Mukerji M, Bhatnagar P, Kannan K, Brahmachari SK. Alu repeat analysis in the complete human genome: trends and variations with respect to genomic composition. *Bioinformatics* 2004;20:813–7.
127. Ram O, Schwartz S, Ast G. Multifactorial interplay controls the splicing profile of Alu-derived exons. *Mol Cell Biol* 2008;28:3513–25.
128. Chen C, Ara T, Gautheret D. Using Alu elements as polyadenylation sites: a case of retroposon exaptation. *Mol Biol Evol* 2009;26:327–34.
129. Payer LM, Steranka JP, Ardeljan D, *et al.* Alu insertion variants alter mRNA splicing. *Nucleic Acids Res* 2019;47:421–31.
130. Mitchell GA, Labuda D, Fontaine G, *et al.* Splice-mediated insertion of an Alu sequence inactivates ornithine delta-aminotransferase: a role for Alu elements in human mutation. *Proc Natl Acad Sci USA* 1991;88:815–9.
131. Ganguly A, Dunbar T, Chen P, Godmilow L, Ganguly T. Exon skipping caused by an intronic insertion of a young Alu Yb9 element leads to severe hemophilia A. *Hum Genet* 2003;113:348–52.

132. Lin L, Jiang P, Shen S, Sato S, Davidson BL, Xing Y. Large-scale analysis of exonized mammalian-wide interspersed repeats in primate genomes. *Hum Mol Genet* 2009;18:2204–14.
133. Sorek R. The birth of new exons: mechanisms and evolutionary consequences. RNA 2007;13:1603–8.
134. Krull M, Brosius J, Schmitz J. Alu-SINE exonization: en route to protein-coding function. *Mol Biol Evol* 2005;22:1702–11.
135. Lev-Maor G, Sorek R, Shomron N, Ast G. The birth of an alternatively spliced exon: 3′ splice-site selection in Alu exons. *Science* 2003; 300:1288–91.
136. Sorek R, Lev-Maor G, Reznik M, *et al.* Minimal conditions for exonization of intronic sequences: 5′ splice site formation in alu exons. *Mol Cell* 2004;14:221–31.
137. Lee JY, Ji Z, Tian B. Phylogenetic analysis of mRNA polyadenylation sites reveals a role of transposable elements in evolution of the 3′-end of genes. *Nucleic Acids Res* 2008;36:5581–90.
138. Bai M, Janicic N, Trivedi S, *et al.* Markedly reduced activity of mutant calcium-sensing receptor with an inserted Alu element from a kindred with familial hypocalciuric hypercalcemia and neonatal severe hyperparathyroidism. *J Clin Invest* 1997;99:1917–25.
139. Cole DE, Yun FH, Wong BY, *et al.* Calcium-sensing receptor mutations and denaturing high performance liquid chromatography. *J Mol Endocrinol* 2009;42:331–9.
140. Janicic N, Pausova Z, Cole DE, Hendy GN. Insertion of an Alu sequence in the Ca(2+)-sensing receptor gene in familial hypocalciuric hypercalcemia and neonatal severe hyperparathyroidism. *Am J Hum Genet* 1995;56:880–6.
141. Arcot SS, Wang Z, Weber JL, Deininger PL, Batzer MA. Alu repeats: a source for the genesis of primate microsatellites. *Genomics* 1995;29:136–44.
142. Chauhan C, Dash D, Grover D, Rajamani J, Mukerji M. Origin and instability of GAA repeats: insights from Alu elements. *J Biomol Struct Dyn* 2002;20:253–63.
143. Lee W, Mun S, Kang K, Hennighausen L, Han K. Genome-wide target site triplication of Alu elements in the human genome. *Gene* 2015;561:283–91.
144. Daniel C, Widmark A, Rigardt D, Ohman M. Editing inducer elements increases A-to-I editing efficiency in the mammalian transcriptome. *Genome Biol* 2017;18:195.

145. Levanon EY, Eisenberg E, Yelin R, *et al.* Systematic identification of abundant A-to-I editing sites in the human transcriptome. *Nat Biotechnol* 2004;22:1001–5.

146. Schaffer AA, Levanon EY. ALU A-to-I RNA editing: Millions of sites and many open questions. *Methods Mol Biol* 2021;2181:149–62.

147. Krestel H, Meier JC. RNA editing and retrotransposons in neurology. *Front Mol Neurosci* 2018;11:163.

148. Vlachogiannis NI, Gatsiou A, Silvestris DA, *et al.* Increased adenosine-to-inosine RNA editing in rheumatoid arthritis. *J Autoimmun* 2020; 106:102329.

149. Lazzari E, Mondala PK, Santos ND, *et al.* Alu-dependent RNA editing of GLI1 promotes malignant regeneration in multiple myeloma. *Nat Commun* 2017;8:1922.

150. Nishikura K. A-to-I editing of coding and non-coding RNAs by ADARs. Nat Rev Mol Cell Biol 2016;17:83–96.

151. Kim Y, Lee JH, Park JE, Cho J, Yi H, Kim VN. PKR is activated by cellular dsRNAs during mitosis and acts as a mitotic regulator. *Genes Dev* 2014;28:1310–22.

152. Chung H, Calis JJA, Wu X, *et al.* Human ADAR1 prevents endogenous RNA from triggering translational shutdown. *Cell* 2018; 172:811–24 e14.

Chapter 5

Retrotransposition as a Cause of Human Disease: An Update

Haig H. Kazazian*

We have known about the existence of human transposons for 32 years since the finding of transposition events in two boys with Hemophilia A[1] and a man with neurofibromatosis I.[2] The boys had LINE-1 (L1) insertions in an exon of their factor VIII gene, and the man had an Alu insertion in an intron of his NFI gene. Over time, we have learned many things about these elements. We know that L1s are autonomous retrotransposons. The roughly 7,000 full-length elements, among 500,000 L1s[3] that are truncated or rearranged, are transcribed into RNA, then reverse transcribed and integrated into the genome in a single step. They supply an element-encoded endonuclease[4] along with a reverse transcriptase[5] to facilitate their insertion. Alu sequences are non-autonomous retrotransposons because they require the reverse transcriptase activity of L1 elements.[6] There are over one million Alus in the human genome, but only a fraction of them can be mobilized. Other non-autonomous sequences that are mobilized by L1s are primate-specific SVA elements,[7-9] processed pseudogenes (copies of cellular mRNAs)[10,11], and small nuclear RNAs,

*Department of Genetic Medicine, Johns Hopkins University School of Medicine, Baltimore, MD 21205, US.

such as U6[12]. L1s are very old, but only the very young ones (<3 My old), the L1Ta subset, are active.[13] Each individual has between 100 and 150 active L1s, but only 5–10 are very active or "hot".[14,15] Since most of these young elements are polymorphic as to presence and sequence within the population,[14,16] there are likely millions of active L1s in the human genomes of the world.[15] However, because human beings have significant defense mechanisms against retrotransposition,[17] the fraction of disease cases caused by retrotransposition events among all the cases of single-gene disease is relatively small, ranging from one in 250[18] to one in 1000.[19]

In 2016, Dustin Hancks and I co-authored an update on the role of retrotransposition in causing human disease.[20] In that paper, we found 124 insertions due to LINE-1 (L1, Alu, SVA, processed pseudogenes, and poly(A) tracks) causing a case of a single-gene disease that had been reported. Now, this report provides an update five years later. I refer the reader to a discussion of those 124 insertions. Moreover, I do not discuss the many somatic L1 insertions that have been found in cancer patients in the present report. (Those data are discussed in the chapter by Ardeljan and Burns.) This report concentrates on retrotransposition events that have caused (or are presumed to have caused) a heritable single-gene disease published since 2015.

However, one point that was discussed in 2015 I will re-emphasize in this review. To my knowledge, four of the 182 heritable retrotransposon insertions mentioned in the current review occurred after fertilization in the early embryo. One L1 insertion was into the X-linked CHM gene causing choroideremia.[21] The somatic nature of the insertion was demonstrated because the mother had only 10% of her white blood cells containing the insertion. She had germinal and somatic mosaicism for the insertion. The second somatic insertion was that of a partial TMF1 gene from chromosome 3 into an intron of the CYBB gene causing chronic granulomatous disease.[22] This unusual processed pseudogene insertion driven by an L1 caused exon skipping in CYBB mRNA processing.

The remaining two cases of somatic, early embryonic insertions were SVAs in the NF1 gene causing neurofibromatosis I.[23] Both of these insertions were associated with large deletions, one of 867 kb,

Table 1. Retrotransposition events causing single-gene disease in humans (2020).

- L1: 39 cases
- Alu: 116 cases
- SVA: 28 cases
- Processed pseudogene: 3 cases
- Poly(A): 4 cases

Total: 190 cases known

and the other of 1Mb. These two events are the largest genomic deletions caused by a *de novo* insertion to date. By sequence analysis, the source elements for both insertions were identified. The first insertion was produced by a full-length SVA from chromosome 6. The second insertion came from an SVA on chromosome 10. The element on chromosome 10 has been associated with other SVA disease-causing insertions and is the likely source element for at least 13 genomic SVAs.[24,25] The somatic nature of these insertions was shown by analyzing the fraction of white blood cells with the deletion. Patient one had the SVA-associated deletion in 93% of her white cells, while the grandmother of patient two had the SVA in 75% of her blood cells.

The number of reported insertion mutations in heritable disease has now risen to 190 (Table 1), with 25 new insertions in hereditary cancers, 10 new insertions in developmental disorders, and 13 new insertions in a large study of possible genetic disease. In this chapter, I will discuss interesting aspects of these 66 new insertions, and comment on what they tell us about human biology.

1. Retrotransposition Events in Large Clinical Studies

Thirty-five of the 66 new insertions causing disease reported since 2015 appeared in a 2017 paper from Myriad Corporation on hereditary cancer.[26] Since fifteen of the 35 insertions were presented in an abstract and were discussed in Ref. [20], only 20 of the insertions are new for this review. In all, 26 cancer-risk, tumor suppressor genes were analyzed, and mobile element insertions were found in 10. In total, 37 retrotransposition insertions were found in 211 people among 10,600 individuals with hereditary cancer studied. The group stated that the overall rate of retrotransposons detected was about 1 in every 325 individuals, with a range depending on the gene involved, of 1 in 110 to 1 in 1165. All of the insertions were germline mutations, and two, an Alu insertion in BRCA2 found in many Portuguese individuals and an SVA insertion in PMS2, had been reported previously by others. There were 6 previously unreported insertions in ATM, 4 in BRCA1, 16 in BRCA2, and 1 or 2 each in APC, CHEK2, MLH1, MSH2, BARD1, and PALB2. It is very interesting that 94% of the insertions were Alus, 33 of 35. All but two of these 33 insertions were identified as Alu Y, the youngest Alu group. The remaining two were merely identified as Alu in the paper. From other studies over many years, the expected fraction of Alu insertions among germline retrotransposition events is ~56%, so this study appears to have found a marked deficiency of L1 and SVA insertions. If Alu insertions accounted for 56% of all insertions in hereditary cancer, then some 25 or so L1 and SVA insertions were missed in this analysis. L1 and SVA insertions are much larger than Alu insertions, so the size difference could account for the deficiency of L1s and SVAs. However, one full-length 6-kb L1 insertion was detected in the study, suggesting that the longer insertion length may not have led to the deficiency of L1s and SVAs in the data. It is also paradoxical that germline insertions in cancer may be nearly 100% Alus, while somatic insertions in cancer are nearly entirely L1s. Another interesting finding in this study involved the ethnicity of the individuals who carried the insertions. There was a deficiency of Americans of European and Ashkenazi Jewish descent, and an excess of Americans of African and Hispanic descent among insertion carriers.

A second study from the Myriad group looked for SVA insertions in hereditary breast cancer. Five likely pathogenic, new SVA insertions were found in APC (2), BRCA1, MLH1, and CHEK2 genes.[27]

A similar excess of Alu insertions was found in a 2020 study from the company Gene Dx, Gaithersburg, MD.[28] They studied, by exome sequencing, over 89,000 individuals representing more than 38,000 cases with suspected single-gene disorders. They made a positive diagnosis in 8700 cases, among which were 13 novel insertions of retrotransposons (frequency = 0.015). Among these insertions, 11 were Alus, one was an L1, and one was an SVA. Of three Alu insertions discussed, all were in exons. All eleven insertions had target site duplications ranging from 9–21 base pairs (bp).

Another important large study published in 2019 searched for retrotransposon insertions in developmental disorders of childhood.[19] In this study, only 11 insertions were found among almost 10,000 children studied, or about 1 insertion in every 1000 affected children. The insertions included 7 Alus, 2 L1s, and 2 processed pseudogenes. Among the 11 insertions, only 4 could be definitely shown to be involved in the cause of the condition. The remaining 7 insertions may have been unrelated to the disorder. However, I have chosen to count them here as potentially disease causing.

2. Retrotransposition in Small Studies and Case Reports

An important paper in the New England Journal of Medicine led by Timothy Yu of Boston Children's Hospital described the diagnosis, molecular characterization, and treatment of a single patient with a serious inborn error of metabolism, Batten's disease or neuronal ceroid lipofuscinosis.[29] A 6 year old with ataxia, blindness, loss of milestones, and seizures was diagnosed with the disease. Molecular analysis found an antisense SVA insertion in intron 6 of the MFSD8 gene. The insertion led to splicing from exon 6 into a cryptic acceptor splice site 119 bp upstream of the SVA and little normal splicing of exon 6 to exon 7 of the gene. The condition is autosomal recessive so the child was a compound heterozygote for a missense mutation derived from the father, and the SVA insertion derived from the mother. Using information obtained from the treatment of spinal muscular atrophy by antisense oligonucleotides (ASOs), Yu and his colleagues tested six ASOs against the cryptic intronic splice site and nearby splice enhancer sequences in the patient's fibroblasts in tissue culture. They found that three ASOs increased normal splicing of exon 6 to exon 7 more than 3-fold. They then picked the best ASO for intrathecal treatment of the patient. After two loading doses, the then seven-year-old patient received an intrathecal dose of ASO every three months. After one year of treatment, her psychological testing showed perhaps slight improvement, but the number and duration of her seizures improved by 50% with the number dropping from 15–30 seizures per day to 0–20 seizures per day. I hope the child continues to improve with longer follow-up. This case is a striking example of individualized medicine, as the mutation is present in only this one patient.

Recently, a very unusual L1 insertion was reported in an intron of the CDKL5 gene in an early infantile epileptic encephalopathy 2 (EIEE2) patient that was likely pathogenic.[30] The total insertion was ~7000 bp. The 3′ segment contained 4272 bp of the 3′ end of an L1Hs and a poly (A) tail. To the 5′ end of the L1Hs was attached 2602 bp of an intron of the PPEF1 gene 230 kb upstream of the L1Hs. The PPEF1 intronic sequence contained ~2000 bp of an old L1 (L1PA5) and an AluSx. The entire ~7-kb insertion was surrounded by a long target site duplication (119 bp) that included and duplicated exon 3 of the CDKL5 gene. This type of complex L1 insertion was seen previously in a tissue culture assay,[31] and likely occurs from the retrotransposing L1 cDNA invading and copying DNA many kb upstream.

Among the remaining 16 retrotransposition insertions in patients, two were X linked recessives; one was a truncated L1 insertion in the first intron of the androgen receptor gene causing a mild case of androgen insensitivity,[32] and the second was an SVA insertion into an exon of the CHM gene causing choroideremia.[33] The latter insertion caused skipping of exon 2 in RNA processing. Three other insertions caused an autosomal dominant condition. Two cases of SVA insertions were found in Lynch syndrome in exons 5 and 3, respectively, of the mismatch repair genes, MSH6 and MSH2.[34] The third insertion was a MAST-2 SVA that accompanied a 37-kb deletion of three exons of the fibrillin (FBN) gene causing a mild case of Marfan syndrome.[35]

Five other insertions were homozygous; four caused autosomal recessive disease and the fifth eliminated the function of a tumor suppressor gene. The first was a full-length L1 insertion in an intron of the SLCA17A gene causing mild Salla disease.[36] The disease was seen in the homozygous offspring of Kurdish parents who were said to be unrelated, but a distant relationship of the parents is suspected because of the rare mutation seen in both of them. A second report was in abstract form in 2015 and was mentioned in Hancks and Kazazian (2016) among known cases at that time. This case involved two families with pyruvate kinase deficiency caused by an Alu insertion into an exon of the PKLR gene. Both patients had nonspherocytic hemolytic anemia affecting their red blood cells and both were the products of consanguineous matings, one Saudi Arabian couple and the second Kuwaiti.[37] The third case was a homozygous insertion of a truncated L1 in an exon of the HACD1 gene causing myopathy. In this case, the parents were consanguineous and Sri Lankan in origin.[38] The fourth insertion was a homozygous deletion of exons 20 and 21 of the WDR66 gene accompanied by an SVA insertion causing multiple malformation of sperm flagellae (MMAF). This deletion–insertion mutation has been in the population of N. Africa between 675 and 1000 years.[39] A fifth case was that of two siblings afflicted with teratoid rhabdoid tumors due to a homozygous SVA insertion into intron 2 of the SMARB1 gene. A single chromosome 22 carried the SVA insertion, but homozygosity was acquired through uniparental isodisomy of the copy of the mother's chromosome 22 carrying the SVA insertion. The insertion caused skipping of exon 2, knocking out this tumor suppressor gene.[40]

Other cases of autosomal recessive conditions were compound heterozygotes, where one allele contained an insertion event and the other contained a common point mutation. One case was an Alu insertion into an allele of the BUB1B gene causing a microcephaly syndrome (PCS/MVA).[41] Another involved an SVA insertion in an exon of the BBS1 gene causing Bardet-Biedl syndrome.[42] The second allele contained a common missense mutation. This SVA insertion has subsequently been observed in eight other Bardet-Biedl patients, but has not been observed in over 10,000 normal individuals.[43] A third case was a germline SVA insertion in an exon of BRCA1, a tumor suppressor gene, in triple negative breast cancer. The second allele was knocked out by loss of heterozygosity (deletion of the second allele of BRCA1), predisposing the patient to breast cancer.[44] A fourth case was an SVA insertion in exon 16 of the BRCA2 gene in a 40-year-old woman with breast cancer.[45] Another autosomal recessive disease due to an L1 insertion produced a GNE myopathy in a Chinese individual. Little information is available for this report, but a different point mutation was likely present on the second allele as only one retrotransposon event was present among 79 mutations found in 45 patients.[46]

As for retrotranposition events in complex conditions, a large study of over 2200 families with an affected child with autism spectrum disorder (ASD) has recently been reported.[47] A number of retrotransposition events potentially related to the basis of the condition were discovered, but only one was considered highly likely. That insertion was a *de novo* Alu insertion in an exon of the CSDE1 gene which had been implicated previously in an ASD individual.[48] Sequencing reads suggested that this insertion might have been somatic as opposed to germline. The phenotype was similar in both cases – intellectual disability, macrocephaly, and vision impairment.

3. Conclusion

Among 66 new retrotransposon insertions likely causing heritable disease reported over the past five years, 44 have come from very large diagnostic studies. These studies were made possible by the widespread use of gene panels for specific types of conditions and exon sequencing. Single case reports of retrotransposition events are still infrequently reported. Sixty percent of the new cases (33/55) were Alu insertions discovered in the three large studies, suggesting that some L1 and SVA insertions were missed, possibly because of their larger size. In small-scale studies or single case reports, L1 and SVA insertions numbered nine and 13, respectively, while only one Alu insertion was found.

References

1. Kazazian HH, Wong C, Youssoufian H, Scott AF, Phillips DG, Antonarakis SE. Haemophilia A resulting from *de novo* insertion of L1 sequences represents a novel mechanism for mutation in man. *Nature* 1988;332:164–6.

2. Wallace MR, Andersen LB, Saulino AM, Gregory PE, Glover TW, Collins FS. A *de novo* Alu insertion results in neurofibromatosis type 1. *Nature* 1991;353:864–6.

3. Lander ES, Linton LM, Birren B, Nusbaum C, Zody MC, Baldwin J, Devon K, Dewar K, Doyle M, FitzHugh W, Funke R, Gage D, Harris K, Heaford A, Howland J, Kann L, Lehoczky J, LeVine R, McEwan P, McKernan K, Meldrim J, Mesirov JP, Miranda C, Morris W, Naylor J, Raymond C, Rosetti M, Santos R, Sheridan A, Sougnez C, *et al.* Initial sequencing and analysis of the human genome. *Nature* 2001;409: 860–921.

4. Feng Q, Moran JV, Kazazian HH, Jr., Boeke JD. Human L1 retrotransposon encodes a conserved endonuclease required for retrotransposition. *Cell* 1996;87:905–16.

5. Mathias SL, Scott AF, Kazazian HH, Boeke JD, Gabriel A. Reverse transcriptase encoded by a human transposable element. *Science* 1991; 254:1808–10.

6. Dewannieux M, Esnault C, Heidmann T. LINE-mediated retrotransposition of marked Alu sequences. Nat Genet 2003;35:41–8.

7. Ostertag EM, Goodier JL, Zhang Y, Kazazian HH, Jr. SVA elements are nonautonomous retrotransposons that cause human disease. *Am J Hum Genet* 2003;73:1444–51.

8. Hancks DC, DC, Goodier JL, Mandal PK, Cheung LE, Kazazian HH, Jr. Retrotransposition of marked SVA elements by human L1s in cultured cells. *Hum Mol Genet* 2011;20:3386–400.

9. Raiz J, Damert A, Chira S, Held U, Klawitter S, Hamdorf M, Löwer J, Strätling WH, Löwer R, Schumann GG. The non-autonomous retrotransposon SVA is trans-mobilized by the human LINE-1 protein machinery. *Nucleic Acids Res* 2012;40:1666–83.

10. Wei W, Gilbert N, Ooi SL, Lawler JF, Ostertag EM, Kazazian HH, Boeke JD, Moran JV. Human L1 retrotransposition: cis preference versus trans complementation. *Mol Cell Biol* 2001;21:1429–39.

11. Esnault C, Maestre J, Heidmann T. Human LINE retrotransposons generate processed pseudogenes. *Nat Genet* 2000;24:363–7.

12. Doucet AJ, Droc G, Siol O, Audoux J, Gilbert N. U6 snRNA pseudogenes: markers of retrotransposition dynamics in mammals. *Mol Biol Evol* 2015;32:1815–32.

13. Boissinot S, Entezam A, Young L, Munson PJ, Furano AV. The insertional history of an active family of L1 retrotransposons in humans. *Genome Res* 2004;14:1221–31.

14. Brouha B, Schustak J, Badge RM, Lutz-Prigge S, Farley AH, Moran JV, Kazazian HH. Hot L1s account for the bulk of retrotransposition in the human population. *Proc Natl Acad Sci* 2003;100:5280–5.

15. Beck CR, Collier P, Macfarlane C, Malig M, Kidd JM, Eichler EE, Badge RM, Moran JV. LINE-1 retrotransposition activity in human genomes. *Cell* 2010;141:1159–70.

16. Seleme MC, Vetter MR, Cordaux R, Bastone L, Batzer MA, Kazazian HH, Jr. Extensive individual variation in L1 retrotransposition capability contributes to human genetic diversity. *Proc Natl Acad Sci USA* 2006; 103:8036–41.

17. Goodier JL, Ostertag EM, Du K, Kazazian HH, Jr. A novel active L1 retrotransposon subfamily in the mouse. *Genome Res* 2001;11:1677–85.

18. Wimmer K, Callens T, Wernstedt A, Messiaen L. The *NF1* gene contains hotspots for L1 endonuclease-dependent *de novo* Insertion. *PLoS Genet* 2011;7:e1002371–11.

19. Gardner EJ, Prigmore E, Gallone G, Danecek P, Samocha KE, Handsaker J, Gerety SS, Ironfield H, Short PJ, Sifrim A, Singh T, Chandler KE, Clement E, Lachlan KL, Prescott K, Rosser E, FitzPatrick DR, Firth HV, Hurles ME. Contribution of retrotransposition to developmental disorders. *Nat Commun* 2019;10:4630.

20. Hancks DC, Kazazian HH, Jr. Roles for retrotransposon insertions in human disease. *Mobile DNA* 2016;7:9.

21. van den Hurk JAJM, Meij IC, del Carmen SM, Kano H, Nikopoulos K, Hoefsloot LH, Sistermans EA, de Wijs IJ, Mukhopadhyay A, Plomp AS, de Jong PTVM, Kazazian HH, Cremers FPM. L1 retrotransposition can occur early in human embryonic development. *Hum Mol Genet* 2007; 16:1587–92.

22. de Boer M, van Leeuwen K, Geissler J, Weemaes CM, van den Berg TK, Kuijpers TW, Warris A, Roos D. Primary immunodeficiency caused by an exonized retroposed gene copy inserted in the CYBB gene. *Hum Mutat* 2014;35:486–96.

23. Vogt J, Bengesser K, Claes KB, Wimmer K, Mautner V-F, van Minkelen R, Legius E, Brems H, Upadhyaya M, Gel JH, Lazaro C, Rosenbaum T,

Bammert S, Messiaen L, Cooper DN, Kehrer-Sawatzki H. SVA retrotransposon insertion-associated deletion represents a novel mutational mechanism underlying large genomic copy number changes with non-recurrent breakpoints. *Genome Biol* 2014;15:1–17.

24. Damert A, Raiz J, Horn AV, Löwer J, Wang H, Xing J, Batzer MA, Löwer R, Schumann GG. 5' transducing SVA retrotransposon groups spread efficiently throughout the human genome. *Genome Res* 2009; 19:1992–20.

25. Hancks DC, Kazazian HH, Jr. SVA retrotransposons: evolution and genetic instability. *Semin Cancer Biol* 2010;20:234–45.

26. Qian Y, Mancini-DiNardo D, Judkins T, Cox HC, Brown K, Elias M, Singh N, Daniels C, Holladay J, Coffee B, Bowles KR, Roa BB. Identification of pathogenic retrotransposon insertions in cancer predisposition genes. *Cancer Genet.* 2017;216–217:160–9.

27. Elias M, Nix P, Pain D. Mancini-DiNardo B, Coffee B, Roa B. Identification and characterization of SVA retroelement insertions through NGS hereditary cancer panel testing and RNA analysis. Abstract 2343 at 2020 American Society of Human Genetics meeting.

28. Torene RI, Galens K, Liu S, Arvai K, Borroto C, Scuffins J, Zhang Z, Friedman B, Sroka H, Heeley J, Beaver E, Clarke L, Neil S, Walia J, Hull D, Juusola J, Retterer K. Mobile element insertion detection in 89,874 clinical exomes. *Genet Med* 2020;22:974–8.

29. Kim J, Hu C, Moufawad El Achkar C, Black LE, Douville J, Larson A, Pendergast MK, Goldkind SF, Lee EA, Kuniholm A, Soucy A, Vaze J, Belur NR, Fredriksen K, Stojkovska I, Tsytsykova A, Armant M, DiDonato RL, Choi J, Cornelissen L, Pereira LM, Augustine EF, Genetti CA, Dies K, Barton B, Williams L, Goodlett BD, Riley BL, Pasternak A, Berry ER, Pflock KA, Chu S, Reed C, Tyndall K, Agrawal PB, Beggs AH, Grant PE, Urion DK, Snyder RO, Waisbren SE, Poduri A, Park PJ, Patterson A, Biffi A, Mazzulli JR, Bodamer O, Berde CB, Yu TW. Patient-customized oligonucleotide therapy for a rare genetic disease. *N Engl J Med* 2019;381:1644–52.

30. Hiatt SM, Lawlor JMJ, Handley LH, Ramaker RC, Rogers BB, Partridge C, Boston LB, Williams M, Plott CB, Jenkins J, David E. Gray DE, James M. Holt JM, Bowling KM, Bebin EM, Grimwood J, Schmutz J, Cooper GM. Long-read genome sequencing for the diagnosis of neurodevelopmental disorders. bioRXives preprint.

31. Gilbert N, Lutz S, Morrish TA, Moran JV. Multiple fates of L1 retrotransposition intermediates in cultured human cells. *Mol Cell Biol* 2005 Sep;25:7780–95

32. Batista RL, Yamaguchi K, Rodrigues ADS, Nishi MY, Goodier JL, Carvalho LR, Domenice S, Costa EMF, Kazazian HH, Mendonca BB. Mobile DNA in endocrinology: LINE-1 retrotransposon causing partial androgen insensitivity syndrome. *J Clin Endocrinol Metab* 2019;104: 6385–90.

33. Jones KD, Radziwon A, Birch DG, MacDonald IM. A novel SVA retrotransposon insertion in the CHM gene results in loss of REP-1 causing choroideremia. *Ophthalmic Genet* 2020;41:341–4.

34. Yamamoto G, Miyabe I, Tanaka, K, Kakuta M, Watanbe M, Kazakami S, Ishida H, Akagi K. SVA retrotransposon insertion in exon of MMR genes results in aberrant RNA splicing and causes Lynch syndrome. *Eur J Hum Genet* 2020 Dec 8. doi: 10.1038/s41431-020-00779-5.

35. Brett M, Korovesis G, Lai AHM, Lim ECP, Tan EC. Intragenic multiexon deletion in the FBN1 gene in a child with mildly dilated aortic sinus: a retrotransposal event. *J Hum Genet* 2017;62:711–5.

36. Tarailo-Graovac M, Drögemöller BI, Wasserman WW, Ross CJ, van den Ouweland AM, Darin N, Kollberg G, van Karnebeek CD, Blomqvist M. Orphanet Identification of a large intronic transposal insertion in SLC17A5 causing sialic acid storage disease. *J Rare Dis* 2017;12:28.

37. Lesmana H, Dyer L, Li X, Denton J, Griffiths J, Chonat S, Seu KG, Heeney MM, Zhang K, Hopkin RJ, Kalfa TA. Alu element insertion in PKLR gene as a novel cause of pyruvate kinase deficiency in Middle Eastern patients. *Hum Mutat* 2018;39:389–93.

38. Al Amrani F, Gorodetsky C, Hazrati LN, Amburgey K, Gonorazky HD, Dowling JJ. Biallelic LINE insertion mutation in HACD1 causing congenital myopathy. *Neurol Genet* 2020;6:e423.

39. Kherraf ZE, Amiri-Yekta A, Dacheux D, Karaouzène T, Coutton C, Christou-Kent M, Martinez G, Landrein N, Le Tanno P, Fourati Ben Mustapha S, Halouani L, Marrakchi O, Makni M, Latrous H, Kharouf M, Pernet-Gallay K, Gourabi H, Robinson DR, Crouzy S, Blum M, Thierry-Mieg N, Touré A, Zouari R, Arnoult C, Bonhivers M, Ray PF. A homozygous ancestral SVA- insertion-mediated deletion in WDR66 induces multiple morphological abnormalities of the sperm flagellum and male infertility. *Am J Hum Genet* 2018;10:400–12.

40. Sabatella M, Mantere T, Waanders E, Neveling K, Mensenkamp A, *et al.* Enlightening the dark matter of the genome: whole genome imaging identifies a germline retrotransposon insertion iSMARCB1 in two siblings with atypical teratoid rhabdoid tumor. Abstract 2304 at 2020 American Society of Human Genetics meeting.

41. Kato M, Kato T, Hosoba E, Ohashi M, Fujisaki M, Ozaki M, Yamaguchi M, Sameshima H, Kurahashi H. PCS-MVA syndrome caused by an Alu insertion in the BUB1B gene. *Hum Genome Var* 2020;4:17021.

42. Tavares E, Tang CY, Vig A, Li S, Billingsley G, Sung W, Vincent A, Thiruvahindrapuram B, Héon E. Retrotransposon insertion as a novel mutational event in Bardet-Biedl syndrome. *Mol Genet Genomic Med* 2019;7:e00521.

43. Delvallée C, Nicaise S, Antin M, Leuvrey AS, Nourisson E, Leitch CC, Kellaris G, Stoetzel C, Geoffroy V, Scheidecker S, Keren B, Depienne C, Klar J, Dahl N, Deleuze JF, Génin E, Redon R, Demurger F, Devriendt K, Mathieu-Dramard M, Poitou-Bernert C, Odent S, Katsanis N, Mandel JL, Davis EE, Dollfus H, Muller J. A BBS1 SVA F retrotransposon insertion is a frequent cause of Bardet-Biedl syndrome. *Clin Genet* 2021;99:318–24.

44. Staaf J, Glodzik D, Bosch A, Vallon-Christersson J, Reuterswärd C, Häkkinen J, Degasperi A, Amarante TD, Saal LH, Hegardt C, Stobart H, Ehinger A, Larsson C, Rydén L, Loman N, Malmberg M, Kvist A, Ehrencrona H, Davies HR, Borg Å, Nik-Zainal S. Whole genome sequencing of triple-negative breast cancers in a population-based clinical study. *Nat Med* 2019;25:1526–33.

45. Deuitch N, Li S-T, Courtney E, Shaw T, Dent R, Tan V, Yackowski L, Torene R, Berkofsky-Fesslers W, Ngeow J. Early-onset breast cancer in a woman with a germline mobile element insertion resulting in BRCA2 disruption: a case report. *Hum Genome Var* 2020;7:24–6.

46. Chen y, Xi J, Zhu W, Lin J, Luo S, Yue D, Cai S, Sun C, Zhao C, Mitsuhashi S, Nishino I, Xu M, Lu J. GNE myopathy in Chinese population: hotspot and novel mutations. *J Human Genet* 2019;64:11–6.

47. Borges-Monroy R, Chu C, Dias C, Choi J, Lee S, Gao Y, Shin T, Park PJ, Christopher A, Walsh CA, Lee EA. Whole-genome analysis of de novo and polymorphic retrotransposon insertions in Autism Spectrum Disorder. bioRxiv preprint doi:http:/doi.org/10.111.1/2021.01.29.428895.

48. Guo H, Li Y, Shen L, Wang T, Jia X, Liu L, Xu T, Ou M, Hoekzema K, Wu H, *et al.* Disruptive variants of CSDE1 associate with autism and interfere with neuronal development and synaptic transmission. *Sci Adv* 2019;5:eaax2166.

Chapter 6

LINE-1 Retrotransposons, Stem Cells, and Human Neurodevelopmental Disorders

Maria Benitez-Guijarro[†,*], Meriem Benkaddour-Boumzaouad[†,*], and Jose L. Garcia-Perez[†,‡]

1. L1 Retrotransposition can Occur in the Early Stages of Human Embryonic Development

During the last two decades, and as knowledge and expertise of stem cell biology has increased, several laboratories questioned the relationship of stem cells with active transposable elements (TEs), and in particular with Long INterspersed Element class 1 (LINE-1 or L1) retrotransposons (Fig. 1). L1s move using a copy and paste mechanism and are the only active autonomous TEs in the human genome.

In 2007, Van den Hurk et al.[1] demonstrated that L1 retrotransposition can occur in the early stages of human embryonic development. While characterizing a patient with a mutation in the CHM gene, it was revealed that this mutation was generated by a de novo L1

---

*contributed equally, listed alphabetically.
†GENYO, Centre for Genomics and Oncological Research: Pfizer/University of Granada/Andalusian Regional Government, PTS Granada, Granada, Spain.
‡MRC-Human Genetics Unit, Institute of Genetics and Cancer, University of Edinburgh, Western General Hospital, Edinburgh, UK.

retrotransposition event. Surprisingly, the insertion was present in a mosaic manner in the patient's mother. Because the mother was a mosaic for this insertion, these data strongly suggested that the retro-transposition event occurred during early embryogenesis of the mother, before the germline was segregated.

Garcia-Perez *et al.* (2007)[2] demonstrated that human embryonic stem cells (hESCs) naturally express endogenous L1 retrotransposons and could support their retrotransposition *in vitro*. Several undiffer-entiated hESC lines were shown to overexpress endogenous L1 RNAs as well as L1 ribonucleoprotein particles (RNPs).[3-5] Further RT-PCR-sequencing analyzes revealed that expressed L1 mRNAs were derived from currently active (e.g., L1Hs) and inactive L1 subfamilies.[2] The authors exploited a reporter-based *in vitro* mobilization assay and detected a low level of retrotransposition in a panel of hESC lines. This suggested that endogenous L1s could move in embryonic stem cells. More recently, Klawitter *et al.* (2016)[6] and Munoz-Lopez *et al.* (2019)[7] demonstrated ongoing mobilization of endogenous active LINE-1s and Alus during the culturing of hESCs and human-induced pluripotent stem cells (hiPSCs) (Fig. 1).

In a recent study, Flasch *et al.* (2019)[8] developed a new method to identify engineered L1 insertions, characterizing >88,000 L1 inser-tions in four human cell lines that mimic biological niches where L1 retrotransposition occurs in humans: HeLa cells, mimicking L1 retro-transposition in cancer[9]; PA-1 human embryonic carcinoma cells and H9 hESCs, mimicking L1 retrotransposition during early embryo-genesis[10]; and hESC-derived neuronal progenitor cells (NPCs), mim-icking L1 retrotransposition in the brain.[11] *De novo* L1 retrotransposition was also explored in a study by Sultana *et al.* (2019), where >1500 L1 integration sites were mapped using a next-generation DNA sequenc-ing (NGS) method.[12] Within this large group of L1 insertions, Flasch *et al.* characterized an estimated 20,000 insertions in hESCs and NPCs, and the authors concluded that the endonuclease activity of L1 and DNA replication dictate integration preferences and promote widespread integration across the human genome (Fig. 1).

Thus, these studies provided *in vivo* and *in vitro* evidence sup-porting that L1 retrotransposition can occur very early during human embryonic development [reviewed in[11]]. Indeed, these studies

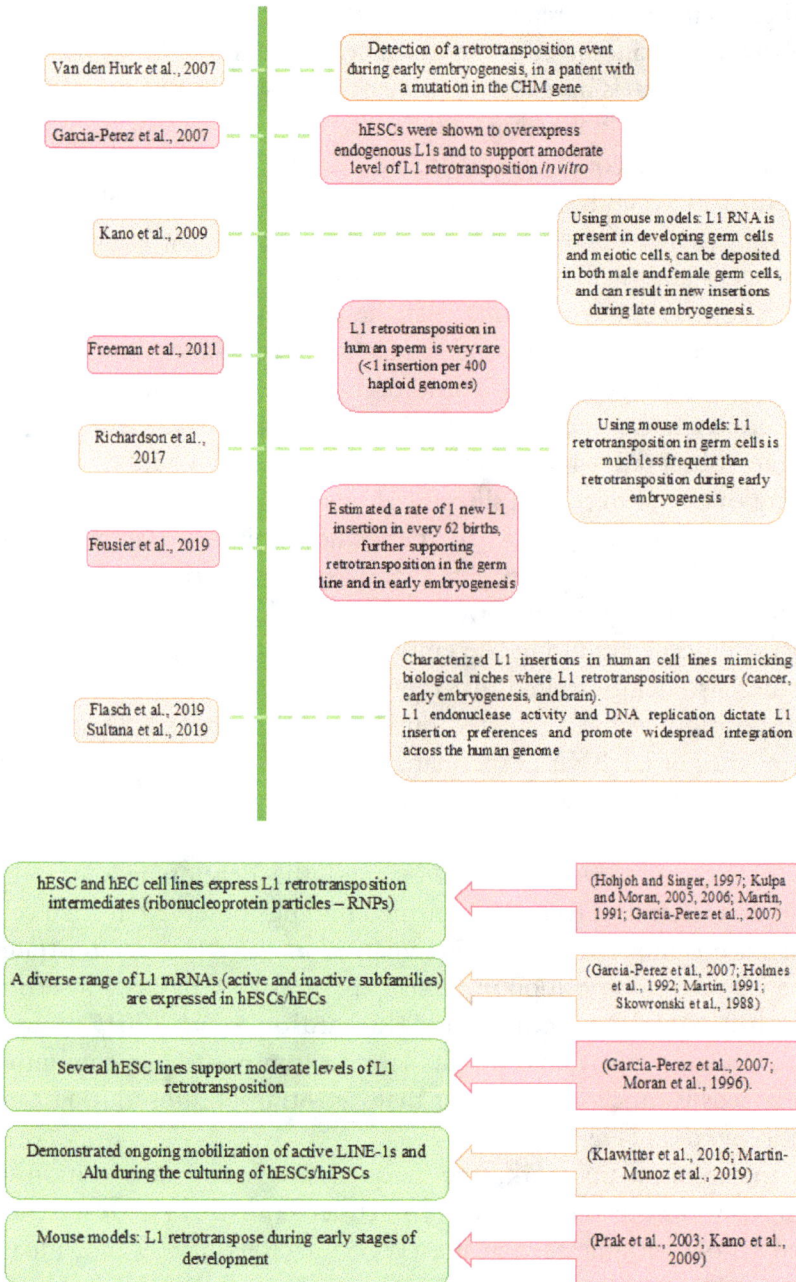

Figure 1. L1 retrotransposition can occur during early stages of human embryonic development.

complemented early reports documenting L1 expression in human embryonic carcinoma cells and germ cell tumors.[13–17]

Mouse models have also been instrumental to learn where L1 can retrotranspose.[18] Pioneering work by the Kazazian lab[19] demonstrated that mouse L1s could retrotranspose during early stages of development. More recently, using similar mouse models of L1 mobilization, the Kazazian lab[20] reported that L1 RNAs are present in developing germ cells and meiotic cells; furthermore, Kano and colleagues provided genetic data suggesting that the L1 RNA is heritable and can be deposited in both male and female germ cells, and that deposited L1 RNAs can generate L1 insertions during late embryogenesis. Indeed, using transgenic mice for human/mouse L1s, Kano and colleagues observed L1 retrotransposition events in >60% of the offspring of carrier mice, and surprisingly also in 9% of the offspring that lacked the L1 transgene.[20] In their study, thirty-five animals were found to be transgenic negative and to present retrotransposition events. A majority of them were non-heritable, suggesting that their insertion occurred during late embryogenesis (Fig. 1). Kano and colleagues also demonstrated that the vast majority of *de novo* L1 retrotransposition events integrated during embryogenesis.[20] Despite the abundance of L1 RNA in spermatogenic cells, a majority of L1 insertions seem to occur after fertilization and may continue throughout development.

The recent *genomics revolution* and the development of NGS methods to study TEs have confirmed findings observed in stem cells and mouse models. Indeed, Freeman *et al.* (2011)[21] demonstrated that L1 retrotransposition in human sperm is very rare (<1 insertion per 400 haploid genomes), and Richardson *et al.* (2017)[22] used mouse pedigrees to demonstrate that retrotransposition in committed germ cells is much less frequent than retrotransposition during early embryogenesis (Fig. 1).

In a follow-up work, the frequency of retrotransposition in humans was studied using NGS. In Feusier *et al.* (2019),[23] researchers used whole genome sequencing (WGS) datasets of 599 individuals, comprising 33 three-generation pedigrees from the Utah Centre d'Etude du Polymorphisme Humain (CEPH), to estimate the rate of retrotransposition for different TEs. Using several pipelines developed to identify mobile element insertions (MEIs) in NGS datasets, such as MELT,[24] RUFUS,[25] and TranSurVeyor,[26] this study estimated a rate of one new insertion of L1 in every 62 births.[23] Consistent with previous studies, Feusier *et al.*, further demonstrated that the majority of L1 insertions accumulated during early embryogenesis. In sum, this study established an accurate rate of L1 retrotransposition, further supporting the ongoing activity of L1 both in the committed germline and during early embryogenesis (Fig. 1).

The above studies have generated a model to explain the evolutionary success of L1 during mammalian evolution. These data further suggest that embryonic stem cells are a physiological model to learn about L1 biology. Similarly, hiPSCs, obtained by reprograming terminally differentiated cells, are very similar to hESCs in their transcriptome and L1 expression profile.[27] Thus, there is a clear connection between L1 and pluripotency, which is supported by several observations: (i) a variety of hESC/hiPSC lines and human embryonic carcinoma (hEC) cell lines are known to naturally overexpress retrotransposition intermediates (L1-RNPs)[2,4,6,28–30]; (ii) several hECs/hESCs/hiPSCs express a constellation of L1 RNAs (from active and inactive subfamilies)[2,4,6,29–33]; and (iii) hECs/hESCs/hiPSCs can support moderate levels of L1 retrotransposition using reporter-based *in vitro* assays.[2,6,30,31,35] From a mechanistic angle, the expression of L1 in pluripotent cells is consistent with the hypomethylation of genomes that occur after fertilization, as L1 expression is controlled by DNA methylation of their promoters.[34,35] From a biological angle, L1s seem to take advantage of the genome-wide hypomethylation that occurs shortly after fertilization[10] to accumulate new insertions, some of which will be heritable in nature.

2. Controlling L1 Activity in Stem Cells: Epigenetic Mechanisms Repress L1 Expression During Early Embryogenesis

While not all early embryonic L1 insertions might pass to the next generation, the genome of the early mammalian embryo should be protected from these mutations/insults. Indeed, parallel to the description of L1 activity during early embryogenesis, several laboratories also analyzed how stem cells might be protected from the mutagenic potential of active TEs. To characterize the "RETROtranscriptome" of hESCs, Macia *et al.* (2011)[36] analyzed which L1 and Alu elements are expressed in undifferentiated hESCs. They observed that, on average, pluripotent cells express 10–15 times more L1 sense RNA than differentiated cells. In addition, they used conventional RT-PCR-sequencing to identify expressed L1s in pluripotent cells, noting that a wide constellation of L1 RNAs are expressed in hEC and hESC cells. Due to their repetitive nature, identifying the genomic location of transcriptionally active L1s is not an easy task. Although it was later found that the activity of the sense/antisense promoters of the L1 5'UTR is not necessarily coupled,[37] Macia *et al.* explored whether the conserved antisense promoter of L1[38] could be used to map expressed L1s in hESCs. They demonstrated that both promoters were active in hEC/hESCs, and next developed a modified 3'RACE protocol to map the 3' end of transcripts initiated from the antisense promoter of L1. They reasoned that a fraction of L1s expressing antisense transcripts might also express the sense L1 RNA. In doing that, they noticed that expressed antisense L1 transcripts were produced from discrete genomic loci, and that a number of these loci were shared among different hESC lines. Thus, not all L1s are expressed to the same level, at least using the antisense promoter of L1 as a proxy. Remarkably, these authors found that a majority of expressed L1s were located within annotated human genes (Fig. 2).[36]

In studies that analyzed the L1 transcriptome in a panel of human somatic tissues, it was observed that retrotransposons are expressed less, on average, than non-repetitive regions of the genome.[39,40] However, in hESCs, expression of L1s is easily detectable, with a very significant enrichment within known genes, relative to their genomic distribution. These data further suggest that there is an apparent relaxation in the control of L1 expression in pluripotent cells. These results are consistent with those of Garcia-Perez *et al.* (2010)[28] where the efficacy of epigenetic silencing and reactivation of *de novo* L1

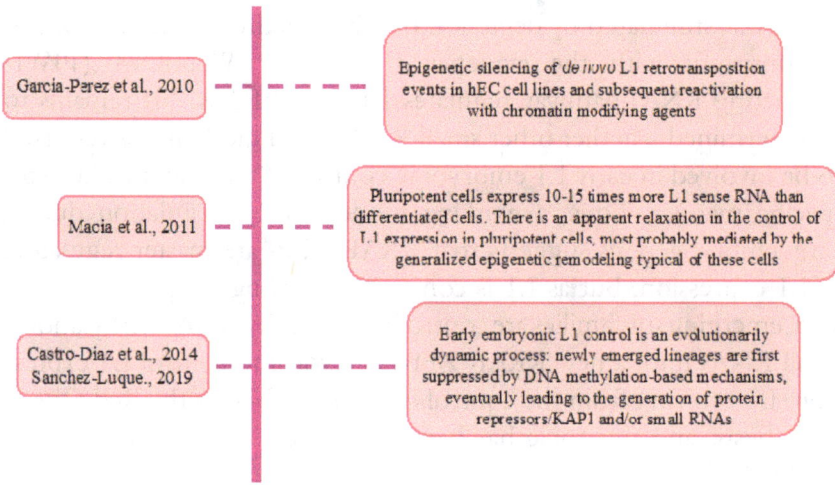

Figure 2. Epigenetic control of L1 expression in pluripotent cells.

retrotransposition insertion in a panel of hEC cell lines was demonstrated (Fig. 2).

Later, a study from the Trono lab demonstrated that the control of L1 expression during early embryogenesis is an evolutionarily dynamic process.[41] These authors demonstrated that evolutionary new emerged lineages of L1 are first suppressed by DNA methylation-based mechanisms, while evolutionary older L1s are suppressed by small RNAs/recruitment of KAP1 (KRAB [Krüppel-associated box domain]-associated protein 1). In fact, KAP1, the master cofactor of KRAB-containing zinc finger proteins (KRAB-ZFPs) previously involved in the restriction of endogenous retroviruses, was found to repress a discrete subset of L1 lineages in hESCs, but lineages that were predicted to have entered the ancestral genome between 26.8 million and 7.6 million years ago. Indeed, they identified a binding site within L1 for repressive KRAB-ZFPs, further suggesting that this rapidly evolving family of proteins would be globally responsible for L1 recognition over evolution (Fig. 2). Further, they observed how the knockdown of KAP1 in hESCs induced the expression of evolutionary older KAP1-linked L1 elements, but not of their younger, human-specific counterparts (L1Hs). However, expression of younger

L1Hs was stimulated by depletion of DNA methyltransferases, which is consistent with the evidence that the PIWI-piRNA (PIWI-interacting RNA) pathway regulates L1Hs in hESCs.[42] It remains to be determined whether other small RNA-based mechanisms reported to be involved in early L1 embryonic control[43-45] also act in a lineage-specific manner. Thus, a working model for L1 regulation during early embryogenesis suggests that KRAB-ZFPs are master regulators of L1 expression, but as L1 is constantly evolving from their repression, emerging subfamilies are controlled mainly by DNA-methylation/ small RNAs until a new KRAB-ZFP can evolve to control L1 expression. L1 evolving from the control of the host follows the Red Queen hypothesis, and this topic has been extensively reviewed in the past (see[10,34,46,47]).

3. The Role of L1 Retrotransposons During Embryonic Development

Because it can negatively impact the host, L1 expression during early embryonic development was thought to represent a weakness of mammalian development that is used by L1 to ensure its evolutionary success. However, recent provocative studies suggest that, on the contrary, L1 could play a role during early embryonic development (Fig. 3).

Jachowicz *et al.* (2017)[48] analyzed the expression of L1 during early stages of development and observed how L1 elements are highly expressed in the early mouse embryo, with maximum expression in the 2-cell (2C) stage and a decrease in expression between the 2- and 8-cell stages. The maximum expression of L1 occurred simultaneous to the moment of cell reprogramming. This observation led these authors to wonder whether the activation of L1 would be relevant for developmental progression. To answer this question, they used engineered nucleases known as TALENs (for transcription activator-like effector nuclease) to extend the levels of L1 expression from the 2-cell stage on, observing that prolonged expression of L1 RNAs interfered with embryo development between the 8- and 16-cell stages, leading to the appearance of morphological abnormalities. Similarly, they used TALENs to inhibit L1 activation before the 2-cell stage, and detected developmental abnormalities. Thus, it seems that the transcriptional activity of L1 after fertilization could be important for embryo development. These observations indicate that activation and silencing of L1 elements play important roles in development. Researching how L1 transcription affects development, they observed that premature silencing of L1 elements decreases accessibility to chromatin, while prolonged activation prevents the gradual compaction of chromatin that occurs naturally in developmental progression.

More recently, Percharde *et al.* (2018)[49] went a step further by considering the possibility that L1 RNAs might play a role in the transcriptional regulation of mouse embryonic stem cells (mESCs). In this work, the authors observed that L1 RNAs were detected at high levels in the nuclei of ESCs, being associated with euchromatin and generally excluded from heterochromatic foci but not from the cytoplasm. A similar location of L1 RNA was also observed in 2-cell mouse embryos and in blastocysts[48] (Fig. 3). The data obtained by FISH indicate that L1 RNAs are necessary for the efficient propagation of ESCs. In addition, they observed that L1 acts to repress MERVL (mouse endogenous retroviral elements of L-type) and the 2C-like transcriptional program in ESCs, since knockdown of L1 induced the conversion of ESCs to a fate similar to 2C. To note, 2C refers to a small subgroup of cells within mESC cultures that exhibits molecular features similar to those of 2-cell mouse embryos.[50] Indeed, it was observed that the reduction of L1 significantly increased the percentage of 2C-like cells showing characteristics as the lack of OCT4 and Nanog proteins, two key pluripotent factors.[50,51] These results indicate that L1 acts to suppress MERVL and the 2C transcriptional program in ESCs (Fig. 3). Finally, studying the mechanism that triggers the repression via L1 RNAs, they discovered that L1 RNAs act as a nuclear scaffold to direct the gene expression program essential for the self-renewal of ESC and the development of the embryo before implantation, as it recruits Nucleolin/Kap1 to repress Dux, a master regulator of zygotic genomic activation (ZGA), and activate rRNA synthesis.

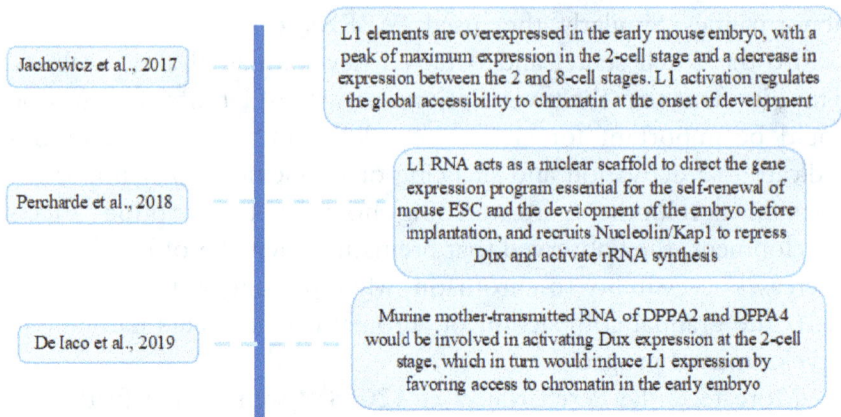

Figure 3.   Does L1 play a role during early embryogenesis?

Later, in De Iaco *et al.* (2019),[52] the first genes of the parental chromosomes that are expressed at the beginning of ZGA, which occurs in the 2C stage in the mouse, were studied. One of the first elements transcribed included the Dux gene, whose product induces a wide range of ZGA genes, and a subset of recent evolutionary L1 retrotransposons that regulate chromatin accessibility in the early embryo. Using mESCs that recapitulate some aspects of ZGA in culture, as they go through a stage similar to 2C when Dux is expressed (i.e., 2C-like), they identified the paralog proteins DPPA2 and DPPA4 as necessary for the activation of Dux and L1 expression in mESCs. Since their coding RNAs are transmitted to the zygote via the mother, these factors are likely to be important upstream mediators of murine ZGA. Thus, the mother-transmitted RNA of DPPA2 and DPPA4 would be involved in activating Dux expression at the 2C stage, which in turn would induce L1 expression by favoring access to chromatin in the early embryo (Fig. 3). Subsequently, L1 would act by inhibiting Dux and activating rRNA synthesis through the recruitment of Nucleolin and Kap1.

These studies are changing the view of L1 as simply selfish DNA, and are paving the way for future research aiming to understand whether L1 RNAs and/or L1 proteins could play additional biological roles. If L1 contribution during embryogenesis will be eventually domesticated is an open question, but it reinforces the concept that TEs are a major driver of genomic innovations.

4. L1 Retrotransposon Activity in Adult Stem Cells

Evidence from cell lines, animal models, and patient characterization has demonstrated that most *de novo* L1 retrotransposition events in humans accumulate during early embryogenesis[1,2,6,20,33] (Fig. 1). Retrotransposition during early embryogenesis has an evolutionary meaning to L1, as a fraction of these insertions would be heritable in nature.[10]

Although with a less clear benefit for the evolutionary success of L1 retrotransposons, several studies have demonstrated that the activity of L1 is not limited to early embryogenesis and germ cells. Indeed, we now know that L1s are expressed and active in at least two non-heritable cellular niches: cancer cells and the brain. In 1992, while characterizing a patient affected with colon cancer, Miki and colleagues found compelling evidence suggesting that L1 retrotransposition could occur in cancer cells.[53] Later, pioneer studies by Moran and colleagues using a cell-based engineered L1 retrotransposition assay demonstrated that L1 insertions can indeed accumulate in tumor cells[54,55] (see the chapter by Ardeljan and Burns and Fig. 4). With the advent of the Genomics Revolution and NGS, our understanding of L1 activity in cancer cells has changed dramatically, and it is well defined now that L1s are expressed and active in many human cancer types explored (recently reviewed in[56]). Because L1 activity in cancer cells has been extensively reviewed, here we will focus on the other main cellular niche where L1 is expressed and active: brain cells (also see the chapter by the Gage lab).

Aiming to better characterize a small group of neuronal cells that can divide in the adult mammalian brain, in 2005, Muotri *et al.*,[11] serendipitously found that L1s are expressed in NPCs isolated from the brain of mice and rats. In this remarkable study, these authors also demonstrated that a human L1 could retrotranspose at a high frequency in rodent NPCs *in vitro*. Using a mouse model of L1

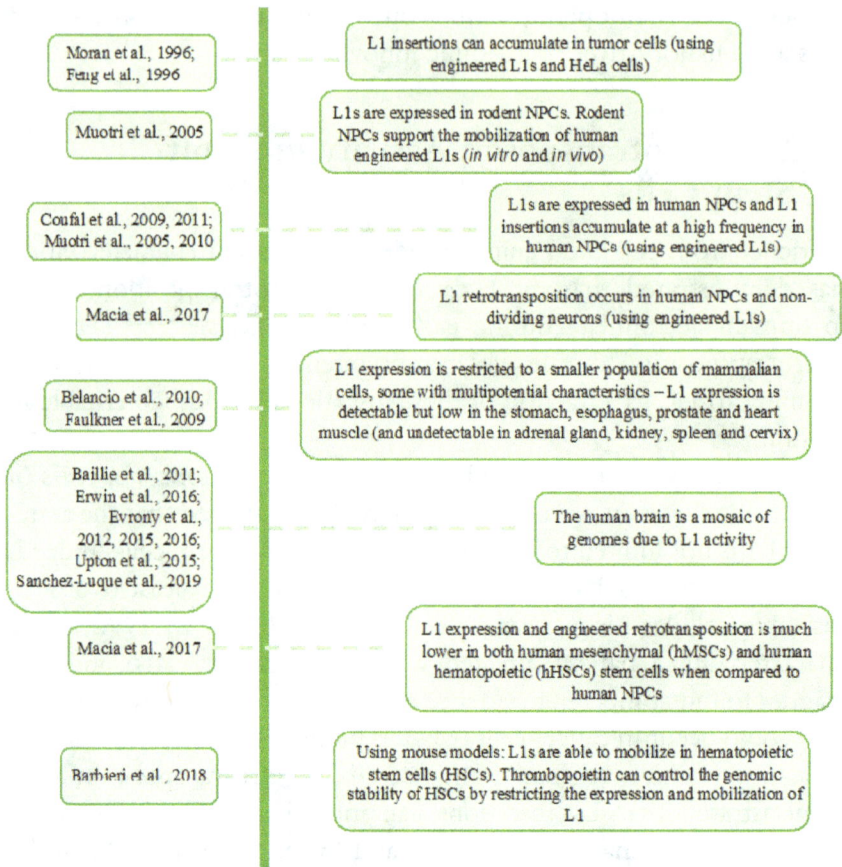

Figure 4.   L1 retrotransposon activity in adult stem cells.

retrotransposition, this study demonstrated for the first time that the rodent brain is a mosaic due to L1 activity (Fig. 4). Human L1s are also expressed and retrotranspose at a high level in human NPCs[57,58] and non-dividing mature neurons.[59] Follow-up studies using NGS approaches further demonstrated ongoing L1 retrotransposition in the human brain,[60–63] although controversy exists for their level of activity in the human brain.[64] The most conservative estimates suggest a rate of 0.6 L1/Alu insertions per neuron, while less conservative estimates suggest >10 insertions per neuron.[60,62,64] The reason for these discrepancies has much to do with technology.[46] No matter what the actual rate of L1 retrotransposition in neurons might be, the human brain contains >10^9 neurons with >10^{15} synapses;[65] Thus, L1-retrotransposition in neuronal cells represents a significant source of genomic variability in the brain.

The discovery of somatic L1 activity in the brain, despite any biological role, also opened a related question: Is this somatic activity of L1 restricted to the brain? Although we know relatively little, several studies have started to explore L1 expression and retrotransposition in other human tissues. Work by Belancio *et al.* (2010)[66] and Faulkner *et al.* (2009) revealed low L1 expression in the human stomach, esophagus, prostate, and heart muscle, while expression in the adrenal gland, kidney, spleen, and cervix was below the limit of detection[39,66] (Fig. 4). More recently, Macia and colleagues[59] took advantage of the differentiation potential of hESCs to compare L1 expression/retrotransposition in isogenic NPCs, mesenchymal, and hematopoietic stem cells (MSCs and HSCs, respectively). Intriguingly, they confirmed a low level of L1 expression in MSCs and HSCs when compared to NPCs, and observed marked differences in L1 retrotransposition.[59] They also used a previously described chimeric vector[67] and observed elevated rates of L1 retrotransposition in NPCs and extremely low retrotransposition levels in MSCs and HSCs.[67] Notably, conclusions gained using hESC-differentiated cell types were similar to previous findings using mouse models of L1 retrotransposition.[11,20]

However, using mouse cells, Barbieri *et al.* (2018)[68] reported that L1, as well as other retroelements, could retrotranspose at a low level in HSCs. Using a mouse model of human L1 retrotransposition, these authors demonstrated that L1 retrotransposition can occur *in vivo* and might be involved in irradiation-induced persistent γH2AX foci and HSC loss of function, a pathway mediated by interferon (INF) activation in response to thrombopoietin. INFs are critical for cellular defense against viruses and are abundantly produced during infections. Whether the same pathway operates in human cells remains to be determined, but this study demonstrated the ability of HSCs to trigger an innate antiviral immune response to a self-renewal cytokine and may represent an important constitutive pathway for resisting the threat of retroelements.

5. L1 and Neurodevelopmental Disorders

Due to the potential role of L1 in brain biology, several studies have analyzed whether L1 deregulation could be involved in neurodevelopmental disorders. While data supporting a role for L1 in neurodegeneration and neurodevelopmental conditions are preliminary, below we discuss some recent studies that are researching in this direction. It has been proposed that epigenetic dysfunctions in the regulation of L1 elements in brain could be involved in neurodegenerative and psychiatric diseases.[69] Consistently, pioneer studies using animal models of L1 retrotransposition showed that environmental factors related to infection or inflammation that disturb early neurodevelopmental processes could increase the number of L1 copies detected in brain.[58,70]

Schizophrenia has been described as a neuro-psychiatric disorder associated with genetic and environmental factors. On the genetic side, there are rare genetic variants that carry a high risk of schizophrenia, although common alleles that carry a moderate risk remain difficult to detect.[71] On this basis, in 2014, Bundo and colleagues[72] reported increased L1 retrotransposition in prefrontal cortex neurons and in hiPSC-derived neurons of patients with schizophrenia that were carriers of the 22q11 deletion (which represents a high genetic risk for schizophrenia). While the validation of *de novo* L1 insertions was very limited in this study, these authors observed that L1 tends to insert preferentially in genes related to synaptic functions and schizophrenia, and proposed that the neuronal genome of schizophrenia patients seems to contain a higher copy number of retrotransposed L1 copies. They concluded that a high level of L1 insertions into neurons may produce genetic disruption in areas related to schizophrenia and synaptic function, leading to an increased susceptibility to schizophrenia[72] (Fig. 5). Despite these provocative findings, these conclusions warrant additional experimental support.

Besides differential accumulation of L1 insertions, other changes influencing the epigenetic regulation of L1 in brain may be a triggering factor which increases the risk of developing schizophrenia. In 2015, Misiak *et al.*,[73] investigated the methylation pattern of L1

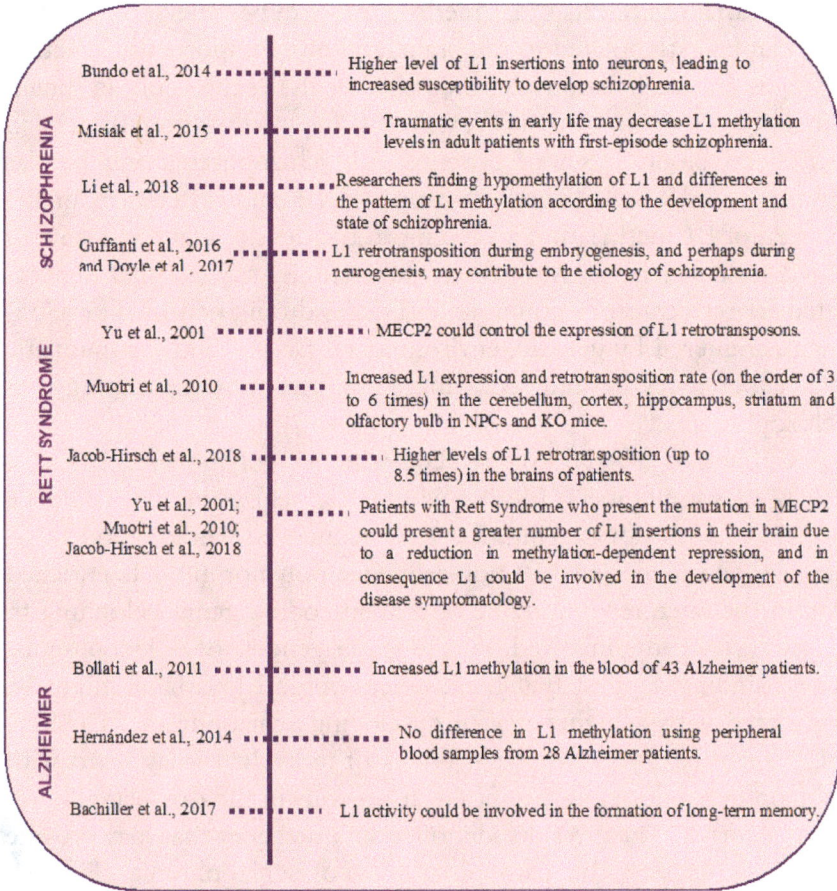

Figure 5. L1 retrotransposon activity in Neurodevelopmental disorders I.

promoters in peripheral blood leukocytes isolated from 48 patients with first-episode schizophrenia or with a history of childhood trauma and compared it to 48 healthy controls without schizophrenia or trauma. Patients with first-episode schizophrenia and childhood trauma had significantly lower L1 methylation compared to first-episode patients without childhood trauma and healthy controls. The results obtained in this study suggest that traumatic events in early life may decrease overall DNA methylation in adult patients with first-episode schizophrenia. However, further research is needed to confirm that the results of L1 methylation with peripheral blood samples can be extrapolated to the brain, and more importantly, whether changes in methylation do indeed alter L1 expression in brain. Similarly, Li e*t al.* (2018)[74] examined the methylation of L1s in peripheral blood DNA of patients with schizophrenia and bipolar disorder, and found hypomethylation of L1 and differences in the pattern of L1 methylation according to the development and state of the disease[75] (Fig. 5). Altogether, these studies suggest that deregulated L1 retrotransposition in critical genes during neuronal development, triggered by genetic, environmental factors, and/or traumatic life experiences, could contribute to the physiopathology of schizophrenia.

Guffanti *et al.* (2016)[74] studied whether polymorphic L1s of a family of six members, four of whom had schizophrenia, could be associated with disease (Fig. 5). They sequenced whole blood DNA samples and identified 110 non-reference polymorphic L1s enriched within the open reading frame of protein-coding genes belonging to pathways already involved in the pathogenesis of schizophrenia. These finding suggest that these polymorphic L1 variants might be associated with a higher risk of developing schizophrenia. Doyle *et al.*,[76] analyzed 36 postmortem brains of individuals diagnosed with schizophrenia, looking specifically at neurons in the dorsolateral prefrontal cortex. These results identified unique genes that may provide new insights into the pathophysiology of schizophrenia (Fig. 5). Indeed, they observed a significant increase of L1 insertions in genes involved in schizophrenia, although these L1 insertions occurred primarily in the germline, with the exception of a possible somatic insertion generated during early development.

Besides schizophrenia, another neurological developmental disorder where L1s seem to play a role is Rett syndrome. Rett syndrome is a rare genetic neurological disorder, which develops around 6–18 months of age and results in late development, intellectual disability, and autistic behavior[77]. This disorder is caused by a mutation in the gene MECP2 (methyl CpG binding protein 2),[78] which is involved in epigenetic regulation. MECP2 binds to CpG dinucleotides, triggering silencing through its interaction with the co-repressor Sin3Ap[79,80] and histone deacetylases. Yu and colleagues, using cultured cells, demonstrated that MECP2 could control the expression of L1 retrotransposons.[81] More recently, Muotri and colleagues[70] further demonstrated how MECP2 could regulate L1 expression *in vivo*, both in human and mice. Comparing the brains of *mecp2* knockout (KO) mice with healthy brains and human NPCs of Rett syndrome, they observed increased L1 expression and retrotransposition rate in the cerebellum, cortex, hippocampus, striatum, and olfactory bulb in NPCs and KO mice. Additionally, using chromatin immunoprecipitation and qPCR, they found higher levels of MECP2 associated with endogenous L1-promoter regions in neural stem cells compared to neurons. Finally, analyzing brains of Rett patients, they observed significantly higher L1 ORF2 copy numbers than in heart tissues of the same patients. Later, Jacob-Hirsch *et al.* (2018)[82] analyzed the complete genome of NPCs from 20 brain samples and 80 other tissue samples from patients with different neurodevelopmental disorders, including patients with Rett syndrome, observing up to 8.5 times higher levels of L1 retrotransposition in the brains of patients than in healthy brains and non-brain tissues. Thus, several pieces of evidence suggest that patients with Rett syndrome could present a greater number of L1 insertions in their brains.

A much more common neurodegenerative disorder linked with L1 is Alzheimer's disease (Fig. 5). Alzheimer's disease is a neurodegenerative disorder that is characterized by the progressive deterioration of memory and cognitive abilities.[83] Bollati and colleagues[84] explored L1 methylation profiles in the blood of 43 Alzheimer's patients using bisulfite pyrosequencing. They reported increased L1 methylation compared to controls, suggesting that the study of L1 methylation may lead to a better understanding of the pathogenesis of Alzheimer's and may help to identify new markers. However, Hernández *et al.* (2014)[85] conducted experiments using peripheral blood samples from 28 patients with Alzheimer's disease, observing no difference in L1 methylation between patients and controls (Fig. 5). Thus, whether L1 could influence cognitive processes and disease remains to be further analyzed. In this context, in 2017, Bachiller *et al.*,[86] found that increased L1 retrotransposition in the hippocampus of adult mice was influenced by neural activation and, by using both pharmacological and genetic approaches in mice, they further concluded that L1 activity could be involved in the formation of long-term memory (Fig. 5). Therefore, available data suggest that LINE-1s could influence memory and cognitive abilities, although more research is warranted.

Besides the abovementioned human neurodevelopmental disorders, there are other diseases related to the brain where a possible role for L1 has been studied, including Aicardi-Goutières syndrome, Ataxia telangiectasia, and Fanconi anemia.

Aicardi-Goutières syndrome (AGS) is a rare genetic disorder that primarily affects the brain and skin, resulting in neurological deficits and often death during early childhood.[87] AGS is characterized by a permanent type I innate immune response induced by INF due to the accumulation of nucleic acids of endogenous origin.[87] Since the original description in the mid-1980s,[88] at least seven genes have been identified as involved in AGS. Notably, these seven genes are all involved in the metabolism of nucleic acids, as well as in the innate immune response to viral infections and endogenous mobile elements.[89] In 2006, four genes associated with AGS were identified: TREX1,[90] and the three subunits of RNAseH2[91]; subsequently, mutations in three additional genes were described in patients showing an AGS-compatible phenotype: SAMHD1,[92] ADAR1,[93] and IFIH1[94] (Fig. 6).

The alteration of only one of these seven genes is capable of triggering AGS. In patients with AGS, a permanent increase in the INF levels from cerebrospinal fluid and serum is observed,[95] along with increased expression of interferon-stimulated genes (ISGs) in peripheral blood.[96] Since no signs of viral infection have been observed in AGS, the role of endogenous mobile elements becomes important. Indeed, most AGS-mutated genes are involved in the innate immune response to viral infections and endogenous mobile elements.[47]

TREX1: In Stetson *et al.* (2008),[97] using engineered LINE-1 retrotransposition assays and cells mutated in TREX1, the authors concluded that TREX1 is a LINE-1 inhibitor (Fig. 6). Stetson *et al.* found 32 times more DNA in the hearts of *trex1* KO mice, and these DNAs

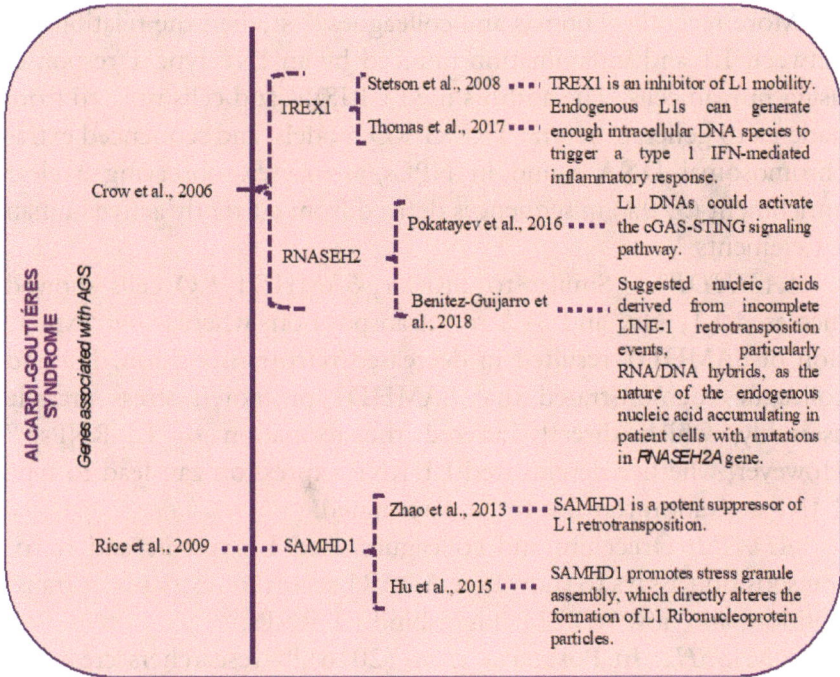

Figure 6.   L1 retrotransposon activity in Aicardi-Goutières Syndrome (AGS).

were enriched in retroelements from LINE, LTR, and SINE subfamilies (more than half were L1 retrotransposons). Using retrotransposition assays, they observed that L1 retrotransposition was reduced >80% upon TREX1 overexpression, concluding that TREX1 is an inhibitor of L1 mobility. In an attempt to test whether this phenotype was reproducible for TREX1 mutations present in patients, they tested loss of function mutants causing recessive AGS (TREX1 V201D and TREX1 R114H mutations) and a catalytic mutant generating autosomal-dominant AGS (TREX1 D200N), and all failed to reduce L1 retrotransposition. Thus, the accumulation of nucleic acids from endogenous L1 retrotransposons could be involved in the development of the physiopathology of AGS (Fig. 6). The key role of L1 in the *trex1* KO mice was further demonstrated by Beck-Engeser *et al.* (2011),[98] where the lethality was rescued by Reverse Transcriptase (RT) inhibitors, although these findings have been recently challenged.[99]

More recently, Thomas and colleagues[100] studied the relationship between L1 and inflammation induced by an INF type 1 response, using human cells. The authors used CRISPR and cells isolated from patients to generate several TREX1 KO models, and sequenced extrachromosomal DNA found in NPCs/astrocytes, observing a clear enrichment (>70%) in sequences derived from currently active human L1 elements.

SAMHD1[101]: Similar to TREX1, SAMHD1 KO cells showed increased L1, Alu, and SVA retrotransposition, whereas overexpression of SAMHD1 resulted in decreased retrotransposition. Hu and colleagues demonstrated that SAMHD1 promoted stress granule assembly, which directly altered the formation of L1-RNPs.[102] However, whether deregulated L1-RNP expression can lead to type 1 INF activation remains to be determined.

ADAR1: Orecchini and colleagues used L1 vectors and retrotransposition assays to show how ADAR1 seems to restrict retrotransposition, and how ADAR1 might bind L1-RNPs.[103]

RNASEH2: In Pokatayev *et al.* (2016),[104] researchers created a mouse model where an RNaseH2 mutation found in AGS patients was introduced; the mutation is a highly conserved residue of the catalytic subunit [rnaseh2a G37S/G37S (G37S)] and has been associated with an early AGS onset. Homozygous *rnaseh2a*-G37S mice were found to be perinatal lethal, and while monitoring the expression levels of ISGs, these researchers established a major role for the cGAS-STING pathway in activating the type 1 INF response. Consistent with a major role for STING in the development of the immune response, a partial rescue of the perinatal lethality was observed in STING KO models. Remarkably, elevated levels of L1 DNAs were observed in the embryos with the G37S mutation, further suggesting that L1 DNAs could activate the cGAS-STING signaling pathway (Fig. 6).

Benitez-Guijarro *et al.* (2018)[105] observed how, contrary to what was expected based on the evidence from TREX1, ADAR1, and SAMHD1, RNaseH2 is necessary for efficient L1 retrotransposition, a result which is consistent with work conducted by Bartsch *et al.* (2017).[106] Using engineered L1 vectors, a collection of RNaseH2A

KO cell lines, and overexpression vectors carrying the cDNA of RNaseH2A containing mutations found in AGS patients (RNASEH2A-G37S and RNASEH2A-E225G), these authors demonstrated that the absence of RNaseH2 activity was associated with a significant decrease in L1 retrotransposition. These results would explain how L1 elements can retrotranspose efficiently without their own RNase H domain, while suggesting that in cases of disease linked to RNaseH2 mutations, the nature of the endogenous nucleic acid accumulating in patient cells could be nucleic acids derived from incomplete LINE-1 retrotransposition events, particularly RNA/DNA hybrids of different lengths. However, further research is needed to confirm this working hypothesis.

The role of these seven AGS genes in controlling the level of endogenous nucleic acids has been analyzed to some level. More recently, mutations in LSM11 and RNU7-1, two genes involved in replication-dependent histone pre-mRNA-processing, have been identified in uncharacterized AGS patients.[107] These data suggest that nuclear histones are essential to suppress the immune response to endogenous nucleic acids.[107]

Although the role of L1 in AGS might be the clearest, there are other neurodegenerative disorders where L1 could also be involved (Fig. 7). Ataxia-telangiectasia (AT) is a neurodegenerative disease with autosomal recessive inheritance caused by mutations in the ATM gene, which encodes for a serine/threonine kinase that acts as a cellular DNA damage sensor and is critical for repairing DNA lesions. AT patients exhibit progressive neurodegeneration, coupled with immunodeficiency, cancer propensity with a variety of inflammatory manifestations, frequent infections, and increased risk of leukemia and lymphoma, with eventual death in the second or third decade of life.[108] ATM is activated by double-stranded breaks,[109] triggering DNA repair mechanisms or p53-mediated apoptosis. Coufal *et al.*,[58] used L1 retrotransposition assays in ATM-depleted cells and systematically observed 2–4-fold increase in L1 retrotransposition. This increase in L1 retrotransposition was also observed using HCT116 colorectal cancer cells deficient in ATM. Notably, these results were recapitulated *in vivo*, using *atm* KO mice that also contained a human

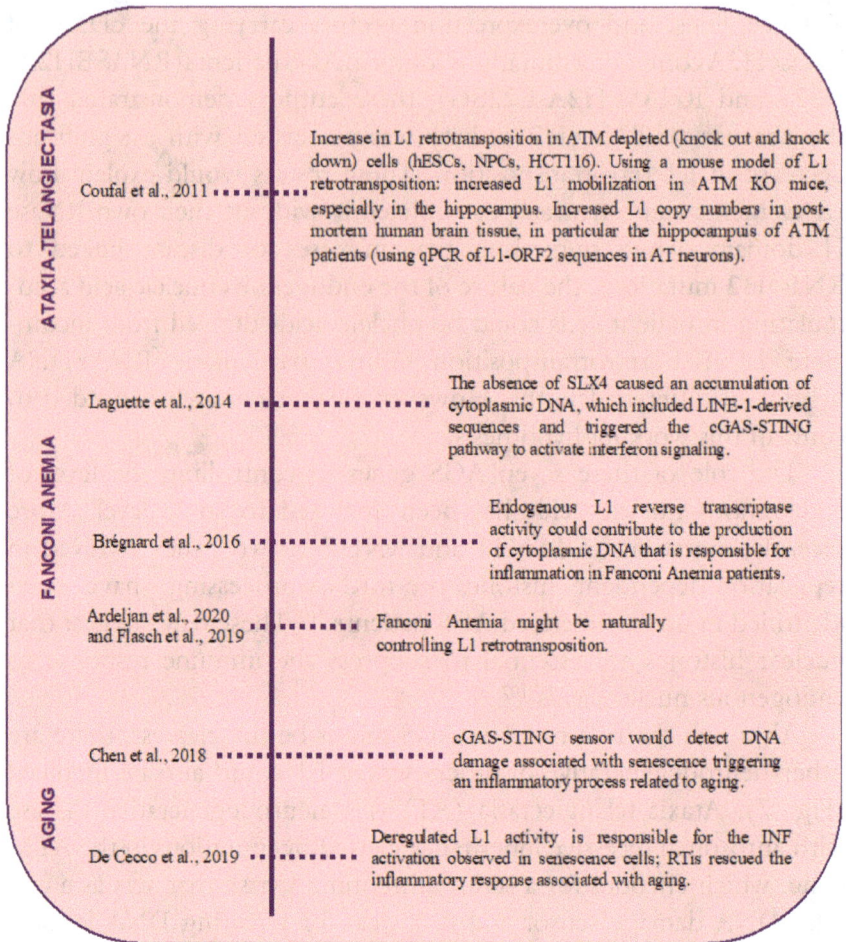

Figure 7. L1 retrotransposon activity in Neurodevelopmental disorders II and other conditions.

L1-transgene (labeled with EGFP), where researchers observed increased L1 retrotransposition especially in the hippocampus. Furthermore, Coufal and colleagues used a sensitive qPCR approach to measure L1 copy numbers in post-mortem human brain tissue, in particular the hippocampus of AT patients, and reported significant higher copy numbers of L1-ORF2 sequences in AT neurons compared to controls[58] (Fig. 7). These results involve L1 in a different neurodevelopmental disease, although it remains to be determined whether increased L1 retrotransposition contributes to the disease process or is simply an indirect consequence in the course of the disease.

More recently, another rare human disease has been associated with deregulated L1 retrotransposition: Fanconi anemia (FA).[110] FA is a recessive genetic disease characterized by bi-allelic mutations in any of the 22 genes identified in the complementation groups of this disease, which constitute the FA pathway involved in DNA repair.[111] The FA pathway is essential for repairing DNA interstrand cross-links (ICLs), and is also critical in the coordination of different DNA repair routes. This leads to increased DNA damage in FA cells, which ultimately results in genomic instability in patients.[112,113]

Laguette *et al.* (2014)[114] demonstrated that a member of the FA pathway, FANCP or SLX4, was directly involved in the repression of HIV-mediated pro-inflammatory signaling, observing that the absence of SLX4 caused an accumulation of cytoplasmic DNA, which included LINE-1-derived sequences and triggered the cGAS-STING pathway to activate INF signaling[114] (Fig. 7). Brégnard *et al.* (2016)[110] reported that the accumulation of cytoplasmic DNA is generated, in part, by increased L1 activity in FANCP mutant cells, and that this increase could be rescued using RT inhibitors. Therefore, they conclude that endogenous L1 reverse transcription could contribute to inflammation in these patients. More recently, additional studies found a connection between L1 and FA,[8,115] further stressing that FA might naturally control L1 retrotransposition (Fig. 7).

Notably, Aicardi Goutières syndrome, Ataxia Telangiectasia, and Fanconi anemia are all associated with inflammation processes related to the activation of an INF type I response, which seems triggered by the DNA sensor STING.[116–118] It is tempting to speculate that under normal conditions there is a tolerance level for transposable element derived intermediates in the cytosol [from LINES, SINEs, and/or endogenous retroviruses], and that when this tolerance level is exceeded, an inflammatory response might develop. Intriguingly, a similar process has been recently associated with aging, where the cGAS-STING sensor would detect DNA damage associated with senescence, triggering an inflammatory process related to aging.[119] Indeed, a very recent report strongly suggest that deregulated L1 activity is responsible for the INF activation observed in senescence cells, and yet again RT inhibitors rescued the inflammatory response associated with aging.[120] While additional research is needed to unveil how L1 deregulation is associated with the abovementioned neurodevelopmental disorders, and perhaps aging, the growing body of evidence suggests that L1s might play a predominant role in these human disorders.

References

1. Van den Hurk JAJM, Meij IC, del Carmen Seleme M, *et al.* L1 retrotransposition can occur early in human embryonic development. *Hum Mol Genet* 2007;16(13):1587–92.
2. Garcia-Perez JL, Marchetto MCN, Muotri AR, *et al.* LINE-1 retrotransposition in human embryonic stem cells. *Hum Mol Genet* 2007;16(13):1569–77.
3. Doucet AJ, Hulme AE, Sahinovic E, *et al.* Characterization of LINE-1 ribonucleoprotein particles. *PLoS Genet* 2010;6(10):e1001150.
4. Kulpa DA, Moran J V. Ribonucleoprotein particle formation is necessary but not sufficient for LINE-1 retrotransposition. *Hum Mol Genet* 2005;14(21):3237–48.
5. Kulpa DA, Moran J V. Cis-preferential LINE-1 reverse transcriptase activity in ribonucleoprotein particles. *Nat Struct Mol Biol* 2006;13(7):655–60.
6. Klawitter S, Fuchs N V, Upton KR, *et al.* Reprogramming triggers endogenous L1 and Alu retrotransposition in human induced pluripotent stem cells. *Nat Commun* 2016;7(1):10286.
7. Muñoz-Lopez M, Vilar R, Philippe C, *et al.* LINE-1 retrotransposition impacts the genome of human pre-implantation embryos and extraembryonic tissues. *bioRxiv* 2019;522623.
8. Flasch DA, Macia Á, Sánchez L, *et al.* Genome-wide de novo L1 retrotransposition connects endonuclease activity with replication. *Cell* 2019;177(4):837–51.
9. Payer LM, Burns KH. Transposable elements in human genetic disease. *Nat Rev Genet* 2019;20(12):760–72.

10. Garcia-Perez JL, Widmann TJ, Adams IR. The impact of transposable elements on mammalian development. *Development* 2016;143(22): 4101–14.

11. Muotri AR, Chu VT, Marchetto MCN, Deng W, Moran J V, Gage FH. Somatic mosaicism in neuronal precursor cells mediated by L1 retrotransposition. *Nature* 2005;435(7044):903–10.

12. Sultana T, van Essen D, Siol O, *et al.* The landscape of L1 retrotransposons in the human genome is shaped by pre-insertion sequence biases and post-insertion selection. *Mol Cell* 2019;74(3):555–70.e7.

13. Bratthauer GL, Fanning TG. Active LINE-1 retrotransposons in human testicular cancer. *Oncogene* 1992;7(3):507–10.

14. Bratthauer GL, Fanning TG. LINE-1 retrotransposon expression in pediatric germ cell tumors. *Cancer* 1993;71(7):2383–6.

15. Ergün S, Buschmann C, Heukeshoven J, *et al.* Cell type-specific expression of LINE-1 open reading frames 1 and 2 in fetal and adult human tissues. *J Biol Chem* 2004;279(26):27753–63.

16. Hohjoh H, Singer MF. Cytoplasmic ribonucleoprotein complexes containing human LINE-1 protein and RNA. *EMBO J* 1996;15(3): 630–9.

17. Skowronski J, Fanning TG, Singer MF. Unit-length line-1 transcripts in human teratocarcinoma cells. *Mol Cell Biol* 1988;8(4):1385–97.

18. Ostertag EM, DeBerardinis RJ, Goodier JL, *et al.* A mouse model of human L1 retrotransposition. *Nat Genet* 2002;32(4):655–60.

19. Prak ETL, Dodson AW, Farkash EA, Kazazian HH. Tracking an embryonic L1 retrotransposition event. *Proc Natl Acad Sci* 2003; 100(4):1832–7.

20. Kano H, Godoy I, Courtney C, *et al.* L1 retrotransposition occurs mainly in embryogenesis and creates somatic mosaicism. *Genes Dev* 2009;23(11):1303–12.

21. Freeman P, Macfarlane C, Collier P, Jeffreys AJ, Badge RM. L1 hybridization enrichment: a method for directly accessing de novo L1 insertions in the human germline. *Hum Mutat* 2011;32(8):978–88.

22. Richardson SR, Gerdes P, Gerhardt DJ, *et al.* Heritable L1 retrotransposition in the mouse primordial germline and early embryo. *Genome Res* 2017;27(8):1395–405.

23. Feusier J, Watkins WS, Thomas J, *et al.* Pedigree-based estimation of human mobile element retrotransposition rates. *Genome Res* 2019; 29(10):1567–77.

24. Gardner EJ, Lam VK, Harris DN, *et al.* The Mobile Element Locator Tool (MELT): population-scale mobile element discovery and biology. *Genome Res* 2017;27(11):1916–29.

25. Ostrander BEP, Butterfield RJ, Pedersen BS, *et al.* Whole-genome analysis for effective clinical diagnosis and gene discovery in early infantile epileptic encephalopathy. *npj Genomic Med* 2018;3(1):22.

26. Rajaby R, Sung W-K. TranSurVeyor: an improved database-free algorithm for finding non-reference transpositions in high-throughput sequencing data. *Nucleic Acids Res* 2018;46(20):e122.

27. Takahashi K, Okita K, Nakagawa M, Yamanaka S. Induction of pluripotent stem cells from fibroblast cultures. *Nat Protoc* 2007;2(12):3081–9.

28. Garcia-Perez JL, Morell M, Scheys JO, *et al.* Epigenetic silencing of engineered L1 retrotransposition events in human embryonic carcinoma cells. *Nature* 2010;466(7307):769–73.

29. Hohjoh H, Singer MF. Sequence-specific single-strand RNA binding protein encoded by the human LINE-1 retrotransposon. *EMBO J* 1997;16(19):6034–43.

30. Martin SL. Ribonucleoprotein particles with LINE-1 RNA in mouse embryonal carcinoma cells. *Mol Cell Biol* 1991;11(9):4804–7.

31. Moran J V, Holmes SE, Naas TP, DeBerardinis RJ, Boeke JD, Kazazian HH. High frequency retrotransposition in cultured mammalian cells. *Cell* 1996;87(5):917–27.

32. Holmes SE, Singer MF, Swergold GD. Studies on p40, the leucine zipper motif-containing protein encoded by the first open reading frame of an active human LINE-1 transposable element. *J Biol Chem* 1992;267(28):19765–8.

33. Wissing S, Munoz-Lopez M, Macia A, *et al.* Reprogramming somatic cells into iPS cells activates LINE-1 retroelement mobility. *Hum Mol Genet* 2012;21(1):208–18.

34. Bestor TH. Cytosine methylation mediates sexual conflict. *Trends Genet* 2003;19(4):185–90.

35. Bourc'his D, Bestor TH. Meiotic catastrophe and retrotransposon reactivation in male germ cells lacking Dnmt3L. *Nature* 2004;431(7004):96–9.

36. Macia A, Munoz-Lopez M, Cortes JL, *et al.* Epigenetic control of retrotransposon expression in human embryonic stem cells. *Mol Cell Biol* 2011;31(2):300–16.

37. Philippe C, Vargas-Landin DB, Doucet AJ, *et al.* Activation of individual L1 retrotransposon instances is restricted to cell-type dependent permissive loci. *Elife* 2016;5:e13926.

38. Speek M. Antisense promoter of human L1 retrotransposon drives transcription of adjacent cellular genes. *Mol Cell Biol* 2001;21(6): 1973–85.

39. Faulkner GJ, Kimura Y, Daub CO, *et al.* The regulated retrotransposon transcriptome of mammalian cells. *Nat Genet* 2009;41(5):563–71.

40. Rangwala SH, Zhang L, Kazazian HH. Many LINE1 elements contribute to the transcriptome of human somatic cells. *Genome Biol* 2009;10(9):1–18.

41. Castro-Diaz N, Ecco G, Coluccio A, *et al.* Evolutionarily dynamic L1 regulation in embryonic stem cells. *Genes Dev* 2014;28(13): 1397–409.

42. Marchetto MCN, Narvaiza I, Denli AM, *et al.* Differential L1 regulation in pluripotent stem cells of humans and apes. *Nature* 2013;503(7477):525–9.

43. Ciaudo C, Jay F, Okamoto I, *et al.* RNAi-dependent and independent control of LINE1 accumulation and mobility in mouse embryonic stem cells. *PLoS Genet* 2013;9(11):e1003791.

44. Fadloun A, Le Gras S, Jost B, *et al.* Chromatin signatures and retrotransposon profiling in mouse embryos reveal regulation of LINE-1 by RNA. *Nat Struct Mol Biol* 2013;20(3):332–8.

45. Heras SR, MacIas S, Plass M, *et al.* The Microprocessor controls the activity of mammalian retrotransposons. *Nat Struct Mol Biol* 2013;20(10):1173–83.

46. Goodier JL. Restricting retrotransposons: a review. *Mob DNA* 2016; 7(16):30.

47. Volkman HE, Stetson and DB. The enemy within: endogenous retroelements and autoimmune disease. *Nat Immunol* 2014;15(5):206–21.

48. Jachowicz JW, Bing X, Pontabry J, Bošković A, Rando OJ, Torres-Padilla M-E. LINE-1 activation after fertilization regulates global chromatin accessibility in the early mouse embryo. *Nat Genet* 2017;(August).

49. Percharde M, Lin CJ, Yin Y, *et al.* A LINE1-Nucleolin partnership regulates early development and ESC identity. *Cell* 2018;174(2): 391–405.

50. Lu F, Zhang Y. Cell totipotency: molecular features, induction, and maintenance. *Natl Sci Rev* 2015;2(2):217–25.

51. Pan G, Thomson JA. Nanog and transcriptional networks in embryonic stem cell pluripotency. *Cell Res* 2007;17(1):42–9.
52. De Iaco A, Coudray A, Duc J, Trono D. DPPA2 and DPPA4 are necessary to establish a 2C-like state in mouse embryonic stem cells. *EMBO Rep* 2019;20(5):1–10.
53. Miki Y, Nishisho I, Horii A, *et al.* Disruption of the APC gene by a retrotransposal insertion of L1 sequence in a colon cancer. *Cancer Res* 1992;52(3):643–5.
54. Moran J V., Holmes SE, Naas TP, DeBerardinis RJ, Boeke JD, Kazazian HH. High frequency retrotransposition in cultured mammalian cells. *Cell* 1996;87(5):917–27.
55. Feng Q, Moran J V., Kazazian HH, Boeke JD. Human L1 retrotransposon encodes a conserved endonuclease required for retrotransposition. *Cell* 1996;87(5):905–16.
56. Burns KH. Transposable elements in cancer. *Nat Rev Cancer* 2017; 17(7):415–24.
57. Coufal NG, Garcia-Perez JL, Peng GE, *et al.* L1 retrotransposition in human neural progenitor cells. *Nature* 2009;460(7259):1127–31.
58. Coufal NG, Garcia-Perez JL, Peng GE, *et al.* Ataxia telangiectasia mutated (ATM) modulates long interspersed element-1 (L1) retrotransposition in human neural stem cells. *Proc Natl Acad Sci* 2011; 108(51):20382–7.
59. Macia A, Widmann TJ, Heras SR, *et al.* Engineered LINE-1 retrotransposition in nondividing human neurons. *Genome Res* 2017; 27(3):335–48.
60. Baillie JK, Barnett MW, Upton KR, *et al.* Somatic retrotransposition alters the genetic landscape of the human brain. *Nature* 2011; 479(7374):534–7.
61. Erwin JA, Paquola ACM, Singer T, *et al.* L1-associated genomic regions are deleted in somatic cells of the healthy human brain. *Nat Neurosci* 2016;19(12):1583–91.
62. Evrony GD, Cai X, Lee E, *et al.* Single-neuron sequencing analysis of l1 retrotransposition and somatic mutation in the human brain. *Cell* 2012;151(3):483–96.
63. Upton KR, Gerhardt DJ, Jesuadian JS, *et al.* Ubiquitous L1 mosaicism in hippocampal neurons. *Cell* 2015;161(2):228–39.
64. Evrony GD, Lee E, Park PJ, Walsh CA. Resolving rates of mutation in the brain using single-neuron genomics. *Elife* 2016;5(e12966).

65. Tang Y, Nyengaard JR, De Groot DMG, Gundersen HJG. Total regional and global number of synapses in the human brain neocortex. *Synapse* 2001;41(3):258–73.
66. Belancio VP, Roy-Engel AM, Pochampally RR, Deininger P. Somatic expression of LINE-1 elements in human tissues. *Nucleic Acids Res* 2010;38(12):3909–22.
67. Kubo S, Seleme M d. C, Soifer HS, *et al.* L1 retrotransposition in nondividing and primary human somatic cells. *Proc Natl Acad Sci* 2006;103(21):8036–41.
68. Barbieri D, Elvira-Matelot E, Pelinski Y, *et al.* Thrombopoietin protects hematopoietic stem cells from retrotransposon-mediated damage by promoting an antiviral response. J Exp Med 2018;215(5):1463–80.
69. Sananbenesi F, Fischer A. The epigenetic bottleneck of neurodegenerative and psychiatric diseases. *Biol Chem* 2009;390(11):1145–53.
70. Muotri AR, Marchetto MCN, Coufal NG, *et al.* L1 retrotransposition in neurons is modulated by MeCP2. *Nature* 2010;468(7322):443–6.
71. Keshavan MS, Nasrallah HA, Tandon R. Moving ahead with the schizophrenia concept: from the elephant to the mouse. *Schizophr Res* 2011;127(1–3):3–13.
72. Bundo M, Toyoshima M, Okada Y, *et al.* Increased L1 retrotransposition in the neuronal genome in schizophrenia. *Neuron* 2014;81(2):306–13.
73. Misiak B, Szmida E, Karpiński P, Loska O, Sąsiadek MM, Frydecka D. Lower LINE-1 methylation in first-episode schizophrenia patients with the history of childhood trauma. *Epigenomics* 2015;7(8): 1275–85.
74. Guffanti G, Gaudi S, Klengel T, *et al.* LINE1 insertions as a genomic risk factor for schizophrenia: preliminary evidence from an affected family. *Am J Med Genet Part B Neuropsychiatr Genet* 2016;171(4): 534–45.
75. Li S, Yang Q, Hou Y, *et al.* Hypomethylation of LINE-1 elements in schizophrenia and bipolar disorder. *J Psychiatr Res* 2018;107(August): 68–72.
76. Doyle GA, Crist RC, Karatas ET, *et al.* Analysis of LINE-1 elements in DNA from postmortem brains of individuals with schizophrenia. *Neuropsychopharmacology* 2017;42(13):2602–11.
77. Guy J, Cheval H, Selfridge J, Bird A. The role of MeCP2 in the brain. *Annu Rev Cell Dev Biol* 2011;27(1):631–52.

78. Amir RE, Van Den Veyver IB, Wan M, Tran CQ, Francke U, Zoghbi HY. Rett syndrome is caused by mutations in X-linked MECP2, encoding methyl- CpG-binding protein 2. *Nat Genet* 1999;23(2): 185–8.

79. Jones PL, Veenstra GJC, Wade PA, *et al.* Methylated DNA and MeCP2 recruit histone deacetylase to repress transcription. *Nat Genet* 1998;19(2):187–91.

80. Nan X, Ng H, Johnson CA, *et al.* Transcriptional repression by the methyl-CpG-binding protein MeCP2 involves a histone deacetylase complex. *Nature* 1998;393(6683):386–9.

81. Yu F, Zingler N, Schumann G, Strätling WH. Methyl-CpG-binding protein 2 represses LINE-1 expression and retrotransposition but not Alu transcription. *Nucleic Acids Res* 2001;29(21):4493–501.

82. Jacob-Hirsch J, Eyal E, Knisbacher BA, *et al.* Whole-genome sequencing reveals principles of brain retrotransposition in neurodevelopmental disorders. *Cell Res* 2018;28(2):187–203.

83. Blennow K, de Leon MJ, Zetterberg H. Alzheimer's disease. *Lancet* 2006;368(9533):387–403.

84. Bollati V, Galimberti D, Pergoli L, *et al.* DNA methylation in repetitive elements and Alzheimer disease. *Brain Behav Immun* 2011;25(6): 1078–83.

85. Rice GI, Kasher PR, Forte GMA, *et al.* Mutations in ADAR1 cause Aicardi-Goutières syndrome associated with a type i interferon signature. *Nat Genet* 2012;44(11):1243–8.

86. Rice GI, del Toro Duany Y, Jenkinson EM, *et al.* Gain-of-function mutations in IFIH1 cause a spectrum of human disease phenotypes associated with upregulated type I interferon signaling. *Nat Genet* 2014;46(5):503–9.

87. Hernández HG, Mahecha MF, Mejía A, Arboleda H, Forero DA. Global long interspersed nuclear element 1 DNA methylation in a Colombian sample of patients with late-onset Alzheimer's disease. *Am J Alzheimers Dis Other Demen* 2014;29(1):50–3.

88. Bachiller S, del-Pozo-Martín Y, Carrión ÁM. L1 retrotransposition alters the hippocampal genomic landscape enabling memory formation. *Brain Behav Immun* 2017;64(2017):65–70.

89. Crow YJ, Rehwinkel J. Aicardi-Goutieres syndrome and related phenotypes: linking nucleic acid metabolism with autoimmunity. *Hum Mol Genet* 2009;18(R2):R130–6.

90. Aicardi J, Goutières F. A Progressive familial encephalopathy in infancy with calcifications of the basal ganglia and chronic cerebrospinal fluid lymphocytosis. *Ann Neurol* 1984;15(1):49–54.

91. Crow YJ, Manel N. Aicardi-Goutières syndrome and the type I interferonopathies. *Nat Rev Immunol* 2015;15(7):429–40.

92. Crow YJ, Hayward BE, Parmar R, *et al.* Mutations in the gene encoding the 3'-5' DNA exonuclease TREX1 cause Aicardi-Goutières syndrome at the AGS1 locus. *Nat Genet* 2006;38(8):917–20.

93. Crow YJ, Leitch A, Hayward BE, *et al.* Mutations in genes encoding ribonuclease H2 subunits cause Aicardi-Goutières syndrome and mimic congenital viral brain infection. *Nat Genet* 2006;38(8):910–6.

94. Rice GI, Bond J, Asipu A, *et al.* Mutations involved in Aicardi-Goutières syndrome implicate SAMHD1 as regulator of the innate immune response. *Nat Genet* 2009;41(7):829–32.

95. Lebon P, Badoual J, Ponsot G, Goutières F, Hémeury-Cukier F, Aicardi J. Intrathecal synthesis of interferon-alpha in infants with progressive familial encephalopathy. *J Neurol Sci* 1988;84(2–3):201–8.

96. Rice GI, Forte GMA, Szynkiewicz M, *et al.* Assessment of interferon-related biomarkers in Aicardi-Goutières syndrome associated with mutations in TREX1, RNASEH2A, RNASEH2B, RNASEH2C, SAMHD1, and ADAR: a case-control study. *Lancet Neurol* 2013; 12(12):1159–69.

97. Stetson DB, Ko JS, Heidmann T, Medzhitov R. Trex1 prevents cell-intrinsic initiation of autoimmunity. *Cell* 2008;134(4):587–98.

98. Beck-Engeser GB, Eilat D, Wabl M. An autoimmune disease prevented by anti-retroviral drugs. *Retrovirology* 2011;8(1):91.

99. Achleitner M, Kleefisch M, Hennig A, *et al.* Lack of Trex1 causes systemic autoimmunity despite the presence of antiretroviral drugs. *J Immunol* 2017;199(7):2261–9.

100. Thomas CA, Tejwani L, Trujillo CA, *et al.* Modeling of TREX1-dependent autoimmune disease using human stem cells highlights L1 accumulation as a source of neuroinflammation. *Cell Stem Cell* 2017;21(3):319–331.e8.

101. Zhao K, Du J, Han X, *et al.* Modulation of LINE-1 and Alu/SVA retrotransposition by Aicardi-Goutières syndrome-related SAMHD1. *Cell Rep* 2013;4(6):1108–15.

102. Hu S, Li J, Xu F, *et al.* SAMHD1 inhibits LINE-1 retrotransposition by promoting stress granule formation. *PLOS Genet* 2015;11(7): e1005367.

103. Orecchini E, Doria M, Antonioni A, *et al.* ADAR1 restricts LINE-1 retrotransposition. *Nucleic Acids Res* 2017;45(1):155–68.

104. Pokatayev V, Hasin N, Chon H, *et al.* RNase H2 catalytic core Aicardi-Goutières syndrome–related mutant invokes cGAS–STING innate immune-sensing pathway in mice. *J Exp Med* 2016;213(3):329–36.

105. Benitez-Guijarro M, Lopez-Ruiz C, Tarnauskaitė Ž, *et al.* RNase H2, mutated in Aicardi-Goutières syndrome, promotes LINE-1 retrotransposition. *EMBO J* 2018;37(15):e98506.

106. Bartsch K, Knittler K, Borowski C, *et al.* Absence of RNase H2 triggers generation of immunogenic micronuclei removed by autophagy. *Hum Mol Genet* 2017;26(20):3960–72.

107. Uggenti C, Lepelley A, Depp M, *et al.* cGAS-mediated induction of type I interferon due to inborn errors of histone pre-mRNA processing. *Nat Genet* 2020;52(12):1364–72.

108. Gatti R, Perlman S. Ataxia-Telangiectasia. Seattle: Gene Reviews; 1993.

109. Shiloh Y. ATM (ataxia telangiectasia mutated): expanding roles in the DNA damage response and cellular homeostasis. *Biochem Soc Trans* 2001;29(6):661.

110. Brégnard C, Guerra J, Déjardin S, Passalacqua F, Benkirane M, Laguette N. Upregulated LINE-1 activity in the Fanconi Anemia cancer susceptibility syndrome leads to spontaneous pro-inflammatory cytokine production. *EBioMedicine* 2016;8:184–94.

111. Wang AT, Smogorzewska A. SnapShot: Fanconi anemia and associated proteins. *Cell* 2015;160(1–2):354-354.e1.

112. Carreau M, Alon N, Bosnoyan-Collins L, Joenje H, Buchwald M. Drug sensitivity spectra in Fanconi anemia lymphoblastoid cell lines of defined complementation groups. *Mutat Res Repair* 1999;435(1):103–9.

113. Kee Y, D'Andrea AD. Molecular pathogenesis and clinical management of Fanconi anemia. *J Clin Invest* 2012;122(11):3799–806.

114. Laguette N, Brégnard C, Hue P, *et al.* Premature activation of the SLX4 complex by Vpr promotes G2/M arrest and escape from innate immune sensing. *Cell* 2014;156(1–2):134–45.

115. Ardeljan D, Steranka JP, Liu C, *et al.* Cell fitness screens reveal a conflict between LINE-1 retrotransposition and DNA replication. *Nat Struct Mol Biol* 2020;27(2):168–78.

116. Härtlova A, Erttmann SF, Raffi FA, *et al.* DNA damage primes the type I interferon system via the cytosolic DNA sensor STING to promote anti-microbial innate immunity. *Immunity* 2015;42(2):332–43.

117. Mackenzie KJ, Carroll P, Lettice L, *et al.* Ribonuclease H2 mutations induce a cGAS/STING-dependent innate immune response. *EMBO J* 2016;35(8):831–44.
118. Zhou R, Xie X, Li X, *et al.* The triggers of the cGAS-STING pathway and the connection with inflammatory and autoimmune diseases. *Infect Genet Evol* 2020;77:104094.
119. Chen K, Liu J, Cao X. cGAS-STING pathway in senescence-related inflammation. *Natl Sci Rev* 2018;5(3):308–10.
120. De Cecco M, Ito T, Petrashen AP, *et al.* L1 drives IFN in senescent cells and promotes age-associated inflammation. *Nature* 2019; 566(7742):73–8.

https://doi.org/10.1142/9789811249228_0007

Chapter 7

Retrotransposition Mechanisms and Host Factors

Siew Loon Ooi*, Kathleen H. Burns[†], and Jef D. Boeke[‡,§]

1. Background

With the advances in genomics research in the past decade, there have been many surprises learned from the human genome and the somewhat simpler genomes of model organisms. One of its greatest enigmas is the preponderance of transposable elements (TEs), which constitute as much as ~70% of our DNA.[1] TEs have been active during evolution, shaping the genomes we see today. Most of the accumulated genetic changes caused by TEs are probably not detrimental, and those insertions that were associated with significant fitness loss have been pruned by the action of purifying selection. While some heritable diseases have been assigned to *de novo* TE insertions or alleles that have become established in small populations,[2] these are

---

*Pramoedya Biointelligence LLC, Woodinville, WA 98077, USA.

[†]Department of Pathology, Dana Farber Cancer Institute, Boston, MA 02215, USA.

[‡]Institute for Systems Genetics, Department of Biochemistry and Molecular Pharmacology, NYU Langone Health, New York, NY 10016, USA.

[§]Department of Biomedical Engineering, NYU Tandon School of Engineering, Brooklyn, NY 11201, USA.

uncommon. In recent years, the broad potential impact of L1 retro-transposons on human disease pathogenesis and aging has been questioned based on their "desilencing" in senescent and cancer cells, as well as certain autoimmune disorders. These primarily manifest in older, post-reproductive individuals and thus, these deleterious consequences of retrotransposon activity have not been subject to the purifying selection alluded to earlier. These observations have kindled a recent interest in anti-retrotransposon pharmacology.

The presence of TE sequences within the human genome was proposed as a way to provide flexibility and potential for a species to evolve and diversify biological processes by McClintock, as well as Britten and Davidson more than 50 years ago.[3,4] These ideas have undergone a renaissance in recent years, with the recognition that TEs may play roles in the gene regulatory network such as in the innate immune functions and in the mouse circadian rhythm enhancers.[5,6–8]

2. Retrotransposon Types and Classification

Retrotransposons are a class of mobile element related to retroviruses, and share the pivotal enzyme reverse transcriptase (RT), producing DNA (termed cDNA) from RNA, which unites and defines a larger universe of retroelements.[7,9,10] (See the chapters by Arkhipova and colleagues as well as by Miller and Le Grice.) There are two main types of retrotransposons: long terminal repeat (LTR) retrotransposons and non-LTR retrotransposons. Like retroviruses, retrotransposons replicate through a cDNA intermediate produced by RT. In contrast to exogenous retroviruses, retrotransposons mostly lack the *env* gene, and indeed, retrotransposon life cycles do not produce infectious particles.

LTR retrotransposons are flanked by direct LTR sequences ranging from 100 bp to more than 5 kbp, but averaging 300–400 bp, and are structured similar to retroviruses. They are categorized into the Ty1-*copia*-like, Ty3-*gypsy*-like, and *BEL-Pao*-like groups (Fig. 1). This categorization is based on sequence similarity and the organization of the genes, in particular the functional domains of the *POL* gene.

Ty1- and Ty3-like elements are broadly distributed and often found in high copy numbers in animal, fungal, protist, and plant genomes. In contrast, *BEL-Pao*-like elements were found in freshwater microscopic invertebrates, notably bdelloid rotifers, and in a wide variety of metazoans.[11-13]

A vast majority of Ty1-*copia*- and Ty3-*gypsy*-like retrotransposons encode two primary translation products corresponding to the retroviral *gag* and *pol* proteins. Interestingly, a variety of the third open reading frame (ORF) has been identified in the *BEL-Pao*-like retrotransposons within rotifers.[14] This third ORF has substantial protein sequence diversity. The yeast Ty1 retrotransposon is one of the most well-studied LTR transposons. Ty1 is the most active and abundant LTR retrotransposon in the *Saccharomyces cerevisiae* reference strain and other yeast strains associated with human activity.[15] Ty1 and its close relatives are widely distributed in the genus *Saccharomyces*,[16] although there is wide variation in the prevalence in individual strains.[17,18]

Non-LTR retrotransposons do encode RT, but despite this similarity, their life cycle is quite distinct from that of the retroviruses and LTR retrotransposons, and these differences are reflected structurally in the lack of the LTRs (Fig. 1). Non-LTR retrotransposons are widely distributed among animals, fungi, and some plants and are especially abundant in mammalian genomes. Human Long Interspersed Element-1 (LINE-1, or L1) is the only autonomous (protein-coding) mobile element active in human cells. L1 elements constitute about 17% of the human genome and are about 6 kb in full length. L1 consists of a 5′ untranslated region with an embedded RNA polymerase II promoter, two open reading frames (ORF1 and ORF2), and a 3′ UTR with a polyadenylation signal (Fig. 1). Both the L1-encoded ORF1 protein (ORF1p) and ORF2p, plus the polyadenylated L1 RNA, are required for its propagation by transposition. ORF1 encodes a chaperone protein with RNA-binding activity, while ORF2 encodes a protein with endonuclease (EN) and RT domains, plus a C-terminal zinc finger domain with unknown activities. A third very short protein (ORF0) was recently described in primate L1 elements which appears dispensable for retrotransposition.[19] The

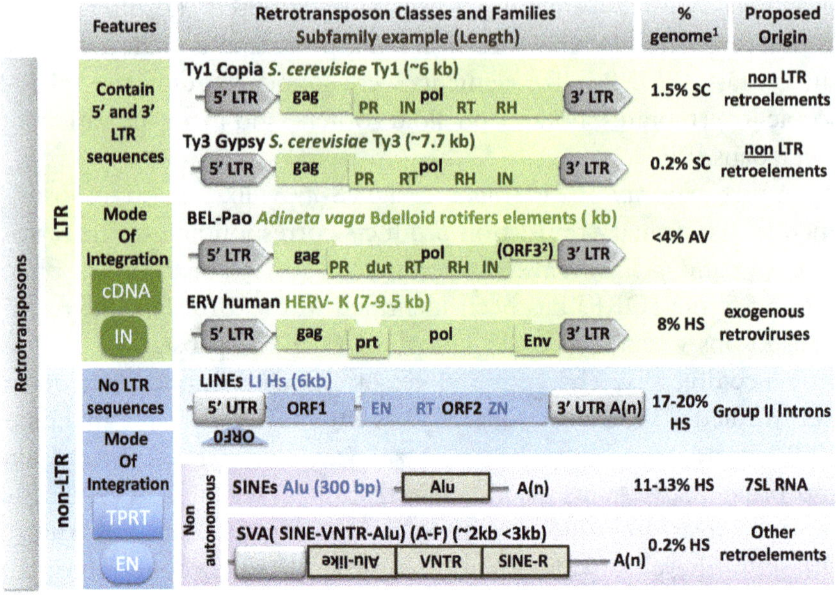

Features	Retrotransposon Classes and Families Subfamily example (Length)	% genome[1]	Proposed Origin

Retrotransposons — LTR

Contain 5′ and 3′ LTR sequences

Ty1 Copia *S. cerevisiae* Ty1 (~6 kb)
5′ LTR — gag — PR IN pol RT RH — 3′ LTR — 1.5% SC — non LTR retroelements

Ty3 Gypsy *S. cerevisiae* Ty3 (~7.7 kb)
5′ LTR — gag — PR RT pol RH IN — 3′ LTR — 0.2% SC — non LTR retroelements

Mode Of Integration — cDNA — IN

BEL-Pao *Adineta vaga* Bdelloid rotifers elements (kb)
5′ LTR — gag — PR dut RT pol RH IN (ORF3?) — 3′ LTR — <4% AV

ERV human HERV-K (7-9.5 kb)
5′ LTR — gag — prt — pol — Env — 3′ LTR — 8% HS — exogenous retroviruses

non-LTR

No LTR sequences

LINEs Li Hs (6kb)
5′ UTR — ORF1 — EN RT ORF2 ZN — 3′ UTR A(n) — 17-20% HS — Group II Introns
ORF0

Mode Of Integration — TPRT — EN

Non autonomous

SINEs Alu (300 bp) — Alu — A(n) — 11-13% HS — 7SL RNA

SVA(SINE-VNTR-Alu) (A-F) (~2kb <3kb)
Alu-like — VNTR — SINE-R — A(n) — 0.2% HS — Other retroelements

Figure 1. Schematic and classification of retrotransposons.

Retrotransposons are classified into two categories: LTR (Long Terminal Repeats) and non-LTR retrotransposons. LTR retrotransposons are flanked by direct repeat sequences termed LTR at both the 5′ and 3′ end. Classes of LTR retrotransposons include Ty1-copia, Ty3 Gypsy, BEL-Pao, and ERV Human. The LTR retrotransposon encodes *GAG* and *POL*; sometimes, a separate PR ORF separates them and sometimes a third ORF (such as *env*-like genes). Retroelements are thought to have evolved independently from diverse origins. LINE1 Hs encode for a third ORF, ORF0 in the antisense direction. Non-LTR retrotransposons lack LTRs, and include LINEs (Long Interspersed Nucleotide Elements), SINEs (Short Interspersed Nucleotide Elements), and SVAs (SINE-VNTR-Alu elements divided into classes A through F). Note that Penelope-like elements could have LTRs, but many copies are 5′ truncated.[119] Furthermore, the LTR sequences do not resemble those of LTR retrotransposons.[120] In the human genome, LINE elements (blue color) are the only autonomous retrotransposons clearly shown to be active. SINEs and SVA elements (in purple color) are non-autonomous, and depend on LINE-1 for their genomic mobilization. The unique features of LTR retrotransposons include the 5′ and 3′ direct repeat LTRs, the presence of cDNA reverse transcribed by RT within the VLP in the cytoplasm, followed by nuclear import of the preintegration complex and retrotransposition of the imported cDNA guided by Integrase (see Fig. 2). In contrast, non-LTR retrotransposons lack LTR sequences, and retrotransposition is guided by a different mechanism termed TPRT, where reverse transcription of the cDNA occurs at the site of integration in the nucleus, guided by the nicking of L1 EN (see **Fig. 2**).[1]

sequence of L1 RNA is unusual in the sense that, unlike most mammalian genes, it lacks a well-conserved intron exon structure, although there are splice sites within human L1.[19,20]

2.1 *LTR retrotransposons*

Genomic studies have revealed incredible variation in TE content. The freshwater invertebrates, rotifers of the class *Bdelloidea*, were found to have a highly diversified repertoire of transposon families.[14] Furthermore, bdelloid rotifer retrotransposon species encode for a third ORF (ORF3), which encodes diverse types of proteins. In the *Adineta vaga* species of rotifer, the LTR retrotransposon families belong to four major lineages (*Vesta*-like, *Juno*-like, *TelKA*-like families, and *Mag* family). Only one lineage (Vesta6c) contains a canonical *env*-like fusion glycoprotein, while the other lineages contain ORF3s with diverse sequences. Members of the Vesta and Juno clades have ORF3s coding for properties including GDSL esterase/lipase, DEDDy exonuclease, and transmembrane domains (TM). GDSL esterase/lipase is a hydrolytic enzyme with broad substrate specificity, including lipids. It has a distinct GDSL sequence motif, in contrast to many lipases which contain the GxSxG motif.[21] Interestingly, esterase activity has previously been identified in ORF1 from the zebrafish *Danio rerio* ZfL2-1 non-LTR retroelement.[22] The authors speculate a role for lipids and membrane targeting in non-LTR retrotransposition. Perhaps ORF1 in this case has Gag-like membrane-targeting capability that could facilitate ribonucleoprotein complex (RNP)

Figure 1. (Figure on facing page) The percent of the genome occupied by the indicated species is specified.[2] Bel-Pao retrotransposons can encode for a third ORF with diverse domains including Env, GDSL esterase/lipase, DEDDy exonuclease, and transmembrane domains. Abbreviations: LTR=Long Terminal Repeat; PR=protease; Gag=group-specific antigen (coat protein) gene; Pol=polymerase; Env=envelope; IN=integrase; RH=RNAse H; UTR=untranslated regions; ORF=open reading frame; EN=endonuclease domain; RT=reverse transcriptase domain; Zn=Zinc Finger; An=A-rich domain; VNTR=variable number target repeats; TPRT=target primed reverse transcription. Gene structures are not drawn to scale.

assembly and/or membrane-mediated processes such as penetration through host membranes during entry and exit.[22] Another ORF3 type has strong homology to RNase D-like DEDDy-type 3′-5′ exonucleases, which may participate in the 3′-end processing of retrotransposon RNA. Even more surprisingly, each of these ORF3s is also associated with different subsets of *Penelope*-like *Athena* retroelement families. Penelope-like retroelements contain endonuclease, and share a common ancestor with telomerase RT, suggesting that this unusual association with *Athena* might point to gene sharing between different groups of retroelements.[14]

2.2 *Non-LTR retrotransposons*

Much interesting progress has been made in the field of non-LTR retrotransposons. This includes the impact of human L1 in cancer and aging, the identification of host factors, and a network of coregulated genes. L1 was found to be cell cycle regulated.[23–26] However, unlike LTR retrotransposons, the diversity of non-LTR retrotransposons has not expanded as dramatically. Also, their limited representation in easily tractable model organisms has limited functional studies, with a few exceptions.[27–29]

3. Retrotransposition Mechanisms

3.1 *LTR retrotransposons and their similarities to retroviruses*

The basic mechanisms of retrotransposition of Ty1 very much mimic that of retroviral replication (Fig. 2; also see the chapter by Miller and Le Grice). Here, we briefly compare the LTR retrotransposon and retroviral life cycles focusing on their many similarities, and on the structure of the LTR sequence at both the DNA and RNA levels (Fig. 1). Keep in mind that retroviruses package two single plus strand-RNA molecules into each virion, but at the DNA level, once it is integrated into the host genome, the virus is called the "provirus" which can subsequently give rise to progeny viruses. LTR retrotransposons share these same nucleic acid forms, the dimer of genomic

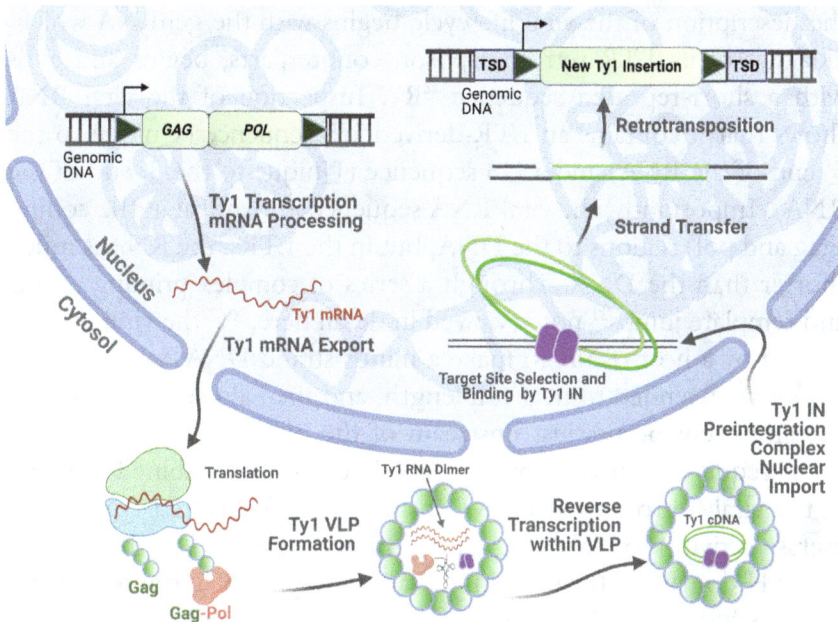

Figure 2. LTR Ty1 life cycle and retrotransposition.
Ty1 RNA is transcribed by RNA polymerase II, and then exported to the cytoplasm. Here, Ty1 RNA is translated into Gag and Gag-Pol proteins, which together with a dimer of Ty1 RNA and tRNA iMet form the virus-like particle (VLP). Ty1 RNA is then reverse transcribed by RT into cDNA within the VLP. tRNA iMet serves as a primer for reverse transcription. Together with Ty1 IN, the newly formed cDNA forms a preintegration complex that is imported into the nucleus. Ty1 IN guides the retrotransposition of Ty1 cDNA into the yeast genome, where it creates a fixed-length 5-bp target site duplication. Figure prepared using BioRender.

RNA (genRNA) in a particle made up of the Gag protein, and referred to as a "VLP" (virus-like particle). The core machinery of these elements consists of the *gag* and *pol* genes which in nearly all cases overlap for some tens of base pairs; a translational frameshifting event produces a Gag-Pol read-through protein at the translational level, at frequencies of ~10–20% of translational cycles.[30] Both Gag, the major capsid protein, and Gag-pol proteins, and their proteolytic cleavage products, are found together with RNA in the VLP. A presumably more "mature" form of the VLP contains the synthesized but unintegrated DNA form that results from reverse transcription, which is subsequently inserted into the genome (Fig. 2). Traditionally, the description of the viral life cycle begins with the genRNA which, like that of its LTR retrotransposon counterparts, begins and ends with a short-repeated sequence "R". Inspection of the viral RNA shows that it contains an LTR-derived U5 sequence (Unique to the 5′ end of the RNA) and a U3 sequence (Unique to the 3′ end of the RNA). Importantly, the viral RNA sequence is identical in the coding (gag and pol) regions to the DNA, but in the LTRs, the RNA is much shorter than the DNA. Through a series of complex priming events and template jumps[31] not reviewed in detail here,[31,32] the viral RNA is primed by a host tRNA to make a minus strand DNA, which is subsequently extended to near full length, and then a second plus strand priming event occurs just upstream of the 3′ LTR at a "polypurine tract" sequence. These steps are carried out by the combined activities of the viral/retrotransposon RT (using its RNA-directed DNA polymerase activity) to make the minus strand. Subsequently, the ribonuclease H domain of the protein degrades the genRNA strand allowing the DNA-directed DNA polymerase activity to produce the plus strand. The product of these reactions is a full-length double-stranded DNA equivalent to a full-length but unintegrated retroviral "provirus". While this is the "text book picture" of this critical DNA intermediate in retrotransposition, which remains associated with the VLP or some partial derivative thereof, the reality is substantially more complex, because the DNA molecules may contain "untidy ends"[33] as well as mutations that occur at substantial rates during the retrotransposition process.[34,35] In a separate step referred to as "integration", this full-length double-stranded DNA is inserted into the genome by the viral/retrotransposon-encoded integrase (IN) protein. The product of this reaction is a full-length element flanked by a "target size duplication" of 5–10 bp (the specific number of bases being defined by the integrase in question) and is considered the *sine qua non* of characterizing an integration site. The following paragraphs summarize some recent findings on LTR retrotransposition mechanisms, and are based on work done on Ty1.

The genRNA of retroviruses and retrotransposons exists in dimeric form in virions and VLPs, respectively.[36] However, little is known about the detailed requirements for Ty1 genRNA dimerization and how packaging occurs in LTR-retrotransposons.

Ty1 Gag was found to be critical for Ty1 RNA stability, nuclear export, and localization.[37] Initiator tRNA-Met, which acts as the primer for the first step in reverse transcription and (−) strand-stop DNA synthesis, is also specifically packaged into VLPs (Fig. 2).[38] RNA sequences required for retrotransposition are located near the 5′ and 3′ termini of the element.[39] The 5′ terminal ~560 nucleotides are known to be sufficient for retrotransposition when co-expressed with a helper element.[40] The 5′-terminal region of Ty1 RNA contains a novel pseudoknot,[41] subsequently identified both *in vivo* and in the virion.[42] A long-range interaction between a 5′ region (CYC5) adjacent to the extended PBS and sequences near the 3′ end of genomic RNA (CYC3) is also critical for efficient reverse transcription and transposition.[43] Surprisingly, evidence suggests that Gag protein must bind Ty1 genomic RNA in the nucleus in order to stabilize it and promote its export to the cytoplasm.[37]

In addition, long non-coding antisense RNAs overlap the 5′-terminal region of the genomic RNA and confer post-translational copy number control.[44] As the copy number increases, the amount of antisense RNA increases. One simple model is that at low copy number, the probability of the two complementary RNAs finding each other is low, but at high copy number, the probability is much higher and expression of Ty1 is tamped down.

A more complex model was proposed on the basis of structural studies of Ty1 RNA, which concluded that antisense RNA was specifically incorporated into VLPs where it might interfere with specific RNA protein binding events required to complete retrotransposition.[45]

Gamache *et al.*[46] found that the 5′ terminus of Ty1 RNA contains RNA motifs required for Ty1 RNA dimerization and packaging. The authors proposed a model wherein a Ty1 RNA kissing complex with two intermolecular kissing-loop interactions initiates dimerization and packaging. The pseudoknot at the 5′ end of Ty1 RNA was found to be important for Ty1 RNA packaging and also for reverse transcription. The pseudoknot consists of the first 326 nucleotides of Ty1 RNA, within which are a 7-bp stem (S1), a 1-nucleotide interhelical loop, and an 8-bp stem (S2), which together delineate two long structured loops. The authors found that mutations that disrupt either pseudoknot stem greatly impacted helper-Ty1-mediated retro-transposition of mini-Ty1 retroelements. However, only mutations in the S2 stem destabilized mini-Ty1 RNA in *cis* and helper-Ty1 RNA in *trans*.

Interestingly, another study found different results for Ty1 RNA dimerization.[42] The interactions of self-complementary PAL1 and PAL2 palindromic sequences localized within the 5′ UTR were found to be essential for Ty1 genRNA dimer formation. Mutations disrupting PAL1-PAL2 complementarity restricted RNA dimerization *in vitro* and Ty1 mobility *in vivo*. In PAL1/PAL2 mutants where dimer formation was reduced, Ty1 RNA could still dimerize via alternative sites. This paper is in contrast to Gamache *et al.*,[46] as the authors were unable to confirm a role for PAL3, tRNA$_i^{Met}$, as well as the recently proposed initial kissing-loop interactions in dimer formation. Another interesting finding from this study is the elucidation of a critical role of Ty1 Gag in RNA dimerization. Ty1 Gag in its mature form binds in the proximity of sequences involved in RNA dimerization and tRNA$_i^{Met}$ annealing. However, the 5′ pseudoknot in Ty1 RNA may constitute a preferred Gag-binding site.

LTR retrotransposons that inhabit small genomes like Ty1 and Ty3 of *Saccharomyces cerevisiae* and Tf1 of *Schizosaccharomyces pombe* show a high degree of targeting in their host genomes.[47,48] Both Ty1 and Ty3, which derive from distinct families of LTR retro-transposons, target tRNA genes, transcribed by RNA polymerase III, using very distinct mechanisms. Tf1, on the other hand, targets RNA Polymerase II-transcribed genes, doing so in a way that does not interfere with transcription of target genes.[48] As a result, these elements avoid insertional inactivation of the genes of their host organisms, which feature tightly clustered genes with very few (and small) introns. Thus, it would appear that purifying selection drove the evolution of these traits.

In contrast, retroviral targeting is far less selective, reflecting the reality that most of the human/mammalian genome is eminently "disrupt-able." Modest trends favoring insertion in transcribed regions are observed in viruses like HIV and Moloney murine leukemia virus, two well-studied examples.[49,50]

3.2 *Non-LTR retrotransposons*

The distinct reverse transcription mechanism used by non-LTR elements like the human L1 element is reflected in major differences in structure relative to the LTR retroelements described in the preceding section (Fig. 2). Most obvious is the lack of anything

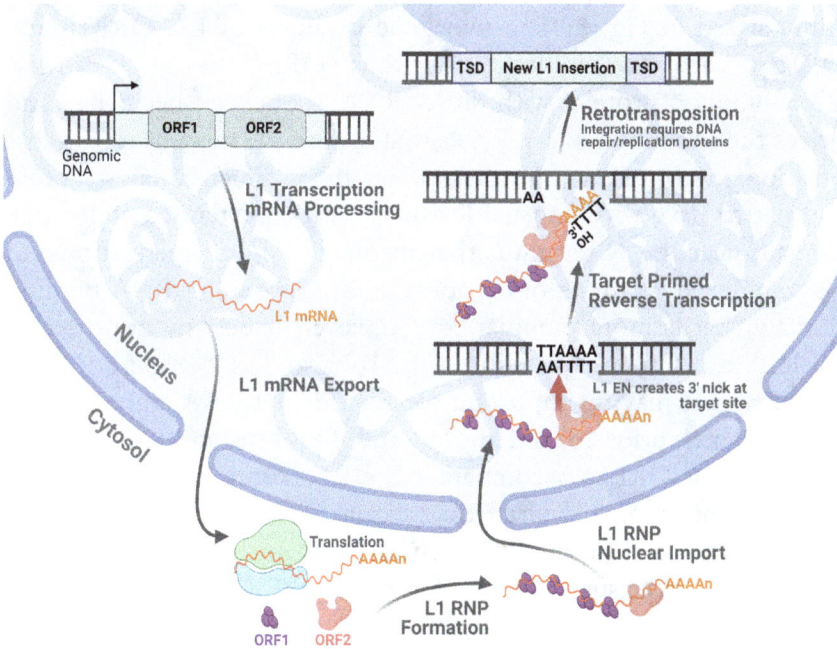

Figure 3.   Non-LTR L1 life cycle and retrotransposition.
L1 RNA is transcribed by RNA polymerase II from L1's own internal promoter, followed by mRNA processing (such as the binding of L1 RNA poly A tail by PABPC1) and exported to the cytoplasm. L1 ORF1 and ORF2 are then translated and form an L1. ORF1 encodes for RNA chaperone activity, and ORF2 encodes for endonuclease and RT activity. The C terminal of ORF2 contains a zinc finger-like domain with unknown activity. L1 RNP then enters the nucleus during mitosis. L1 ORF2 EN creates a 3' OH nick at its preferred genomic target site (5'TTTT/AA) to prime reverse transcription, in a mechanism termed as target-primed reverse transcription (TPRT). This is then followed by first and second strand synthesis of the L1 DNA, followed by its resolution and integration into the genome that most likely involves DNA repair and/or replication factors during S phase. L1 retrotransposition creates a variable length target duplication site, a typical characteristic of most non-LTR retrotransposons. Figure prepared in BioRender.

remotely resembling an LTR at the ends, nor is there any kind of repeat sequence defining the ends of the non-LTR elements. Instead, what we see is what looks very much like the features of an unusual kind of messenger RNA that has been "pasted" into the genome.

The life cycle of human L1, aspects of its core biochemical reaction, and target primed reverse transcription (TPRT, Fig. 3) are summarized in Fig. 3. The two gene products of L1, ORF1p and ORF2p, each have multiple functions. ORF1p forms a trimeric RNA-binding protein with little RNA sequence specificity, and drives the formation of L1-RNPs that apparently serve as retrotransposition intermediates. Studies of its three-dimensional structure show it to have very unusual features, with an extended coiled coil domain near the N-terminus that mediated homotrimerization and is complexed with chloride ions, and a separate RNA-binding domain that is more globular.[51,52] However, like many non-specific nucleic acid-binding proteins, ORF1p also has "chaperone"-like activities that may be important for remodeling RNA and/or DNA intermediates prior to and during the TPRT reaction. Evidence[52] suggests that ORF1p accompanies the RNPs into the nucleus, but then is either rapidly degraded or exported.[53,54] ORF2p is both an endonuclease and a RT and is thought to nick target DNA at a degenerate consensus sequence that matches TTTT/AA, nicking on the "bottom" strand of the target.[55,56] The endonuclease's 3D structure echoes its sequence similarity to its relatives, the AP endonucleases, and other members of the DNAse I family.[56] There is as of yet no structure for the RT, although structures of a related enzyme, telomerase, have been solved.[57] The first stage of TPRT is thought to proceed as outlined in Fig. 3, explaining how the "minus strand" of L1 is synthesized. However, the mechanism of the "second nick" and plus strand synthesis is not fully worked out. Recent studies suggest that the TPRT process occurs during the S phase,[23–26] and there are multiple studies implicating components of the replication machinery as host factors in L1 replication, suggesting a complex relationship with replication fork restart machinery, discussed in a later section.

4. Host Factors

4.1 *Background*

The relationship between TEs and their hosts has often been likened to an "arms race"[58-61] with its roots in the "Red Queen hypothesis"[62] based on the Red Queen who famously told Alice in *Through the Looking Glass,* [63]"Now, *here,* you see, it takes all the running you can do, to keep in the same place". L. Van Valen called it the "Red Queen" hypothesis because species have to "run" or evolve in order to maintain the same relationship, or else go extinct. Retrotransposons, just like other mobile elements and viruses, require interactions with the host factors for their ongoing existence. But, the hosts have developed mechanisms to mitigate potentially harmful effects caused by L1, including the development of active mechanisms to shut down retrotransposition. Many of the most interesting known molecular mechanisms of gene control are likely to have their origins in the evolution of "host defense/coping mechanisms"; examples include the formation of the nuclear envelope at the dawn of eukaryotes,[64] DNA and histone modifications, and the evolution of pre-mRNA splicing machinery. With current advances in genomics and proteomics capabilities, dozens and dozens of host factors have been discovered through diverse functional screens, including traditional forward genetic screens (siRNA, shRNA, CRISPR/CAS9) and protein interactomics screens, to name a few.[23,54,65-67]

4.2 *Assays for retrotransposition*

Nearly all of the "perturbational" screens alluded to above depend on one or more retrotransposition assays. In the constructs used in these assays, the retrotransposon sequence is tagged with what is sometimes referred to as a "retrotransposition reporter" (Fig. 4). In these elements, a functional copy of the element is tagged in a non-essential, non-coding portion of the element, usually just 3′ to the *pol* gene. The reporter gene, complete with promoter and polyadenylation signal, is inserted "antisense", i.e., in the opposite orientation to that of the element itself. Finally, an intron is inserted into the reporter so as

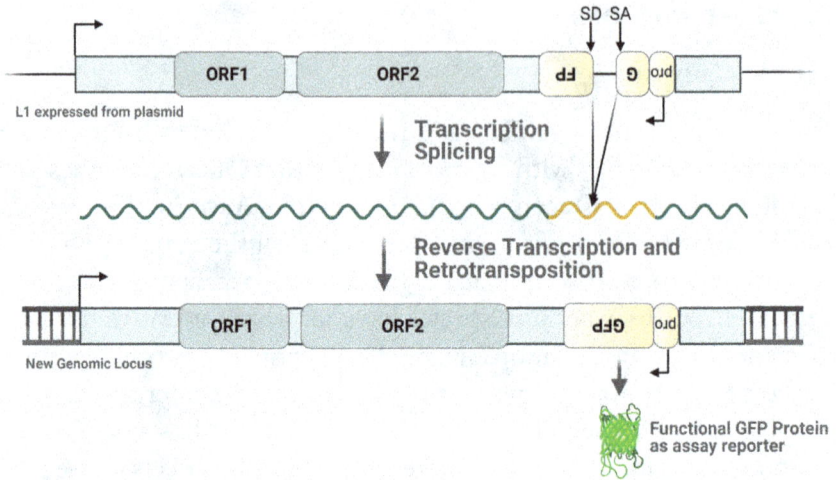

Figure 4.   Basic principle of retrotransposition assays.
Retrotransposition assays, such as one illustrated here for L1, have been well developed in diverse experimental models, and they all have one universal requirement: the demonstration that the identified retrotransposition event occurs via a cDNA intermediate dependent on RT activity. The retrotransposon sequence with a "retrotransposition reporter" such as GFP-AI (artificial intron). In these elements, a functional copy of the element is tagged in a non-essential, non-coding portion of the element, usually at the 3′ end. The reporter gene is inserted "antisense", in the opposite orientation, to that of the element itself, complete with a minimal promoter. An artificial intron is inserted in the "sense orientation" into the reporter so as to disrupt the reporter gene's coding region. This configuration requires that (i) the "sense" intron be spliced out from the "antisense" GFP coding region within the "sense" main L1 transcript, (ii) that cDNA is synthesized from this spliced transcript, and (iii) that only transcription from the minimal promoter of this cDNA, subsequent to its integration, would produce a functional GFP reporter. Thus, if and only if a full cycle of retrotransposition occurs is the retrotransposition reporter expression activated.

to disrupt the reporter gene's coding region; it is inserted in the "sense" orientation so that it will be removed from the retrotransposon transcript, but not the reporter (antisense) transcript. Thus, if and only if a complete retrotransposition occurs is the retrotransposition reporter expression activated. Retrotransposition reporters can be auxotrophic markers (popular in yeast), drug resistance markers (popular in mammalian cells), or fluorescent proteins for single-cell assays or luciferase for the ultimate in sensitivity.

5. Ty1 Host Factors

5.1 *Priming factors*

The sequence of the primer binding site in Ty RNA suggested that the 3′ end of the initiator methionine tRNA might bind there by Watson Crick base pairing. There was a precedent for this with retroviruses, but the initiator tRNA had never been seen to be used previously and the length of the complementarity was always 18 bp in the viruses and in Ty1 it was merely 10 bp long, leaving open the possibility of other mechanisms. Chapman *et al.*[38] engineered both the primer binding site and the tRNA gene itself to produce complementary sets of 5 base changes in the 10 nt complementary sequences. Changes to either the primer binding site or to the tRNA gene completely eliminated retrotransposition, but the combination of the two restored it to near wild-type levels. Primer binding site complementarity to this tRNA and many others has been observed in the vast majority of LTR retrotransposons. An interesting exception is typified by Tf1, an element from *Schizosaccharomyces pombe*, which uses a fragment derived from the 5′ end of Tf1's own mRNA as the primer instead, freeing this element (and related elements) of a dependence on a potentially "troublesome" host factor.[68,69]

5.2 *Integration specificity*

It has long been known that Ty1 shows targeting to tRNA genes. Early work showed that the tRNA gene needed to be transcriptionally active in order to serve as a target, and that the important signal determining insertion upstream of tRNA was the tRNA itself and not the flanking sequence.[70] More recent work using deep sequencing showed that all tRNA genes and other RNA polymerase III (pol III) genes serve as targets.[71,72]

The Ty1-related element Ty5 has also served as an excellent model for targeting specificity. Work by the Voytas lab has shown that the C-terminus of its integrase binds to a specific component of silent chromatin, Sir4, and that a small peptide sequence is sufficient to direct this interaction.[73,74] Thus, Sir4 is a very clear-cut host factor for integration.

The integration of the Ty1 LTR retrotransposon upstream of Pol III-transcribed genes was shown to require interaction between the AC40 subunit of RNA polymerase III RNP complex and Ty1 integrase (IN1).[75] A short amino acid sequence, termed the Ty1 targeting domain (TD1), residing within the bipartite nuclear localization signal (bNLS) sequence of Ty1 IN1 is[76] required for protein interaction between Ty1 Integrase and the AC40 subunit of pol III, also known as the tethering factor for Ty1 integration.[75]

The target specificity and function of TD1 was further confirmed when the authors replaced the Ty5 retrotransposon targeting sequence by the bNLS of Ty1 integrase. Ty5 integrase, IN5, normally interacts with Sir4 and targets pol III genes at subtelomeric regions. Interestingly, when the authors replaced the Ty5 retrotransposon targeting domain (TD5) with the identified TD1, Ty5 integration was redirected from subtelomeric regions to Pol III-transcribed gene loci preferred by Ty1 TD1.[75] This result showed that the Ty1 IN1 bNLS sequence is both required and sufficient to confer integration site specificity on Ty1 and Ty5 retrotransposons. Though Ty1 IN1 is also recruited to Pol I-transcribed genes through its interaction with AC40, Pol I-transcribed genes are poor targets of Ty1 integration. Ty1 TD1 alone is insufficient to recruit Ty1 integration into pol I genes. It seems that another factor is also required. First, Ty1 IN1 interacts directly with AC40 in the absence of other yeast proteins. Mutations in TD1 that reduce the interaction with AC40 abolish Ty1 IN1 association with both Pol I and Pol III transcription complexes. Mutations in bNLS do not impact the frequency of Ty1 retromobility; instead, they decrease the recruitment of Ty1 IN1 to Pol III-transcribed genes and the subsequent integration of Ty1 at these loci. Furthermore, mutations in Ty1 IN1 bNLS induce the same changes in the Ty1 integration profile as observed in the AC40sp loss-of-interaction mutant.[75]

5.3 *Other host factor interactions*

One of the first screens for Ty1 host factors identified 101 host factors.[77] Thus, it was no surprise that numerous additional factors proved to play roles in the regulation and mechanism of retrotransposition. Some interesting recently described Ty1 host factor interactions include between Ty elements and nuclear pore components,[78,79] effects of the mediator complex influencing choice between synthesizing Ty1 v Ty1i (Ty1 inhibitor) RNA,[80,81] and requirement for endoplasmic reticulum localization of Ty1 Gag protein synthesis/stability.[82]

6. LINE-1 Host Factors

6.1 *L1 Background*

L1 is upregulated in more than half of cancers, and has been associated with other disease phenotypes in both germline and somatic cells.[2,83,84] Evidence of L1 activity and transposition has been extensively demonstrated from evolutionary studies.[34,85,86] Recent progress in L1 host factor studies shows multiple pathways used by host factors to restrict L1 activity throughout its life cycle, ranging from epigenetic silencing, transcriptional repression, translation control, and the nuclear import of L1 RNP, to finally, TPRT-mediated integration into the genome. As with other TEs, there is evidence for an intense arms race or competition between L1 and host factors to prevent L1 retrotransposition.[87–89]

Progress in recent years further elucidated how host factors interact and restrict L1, and sometimes are used for retrotransposition. Diverse studies all point to L1 retrotransposition occurring during DNA replication in S phase,[53,54,66] and perhaps by exploiting stalled replication forks.[23,24] This process involves interaction with a broad range of DNA replication and repair factors. This knowledge is helpful to elucidate L1 biology in carcinogenesis and provides opportunities to further explore L1 for therapeutic purposes.[23,24,90]

Another emerging theme is the connection between the host factor pathway recognizing L1 and innate immunity via the Type I interferon signaling pathway.[24,91]

6.2 *Transcriptional regulation and mRNA processing*

Several studies further establish epigenetic silencing, including via HUSH/MORC2 as a way to restrict L1 retrotransposition. In a study where CRISPR/CAS was used to screen two human cell lines (HeLa and K562 cells) for host factors involved in L1 retrotransposition,[65] ~140 human host factors were identified, including those involved in chromatin or transcriptional regulation, DNA damage or repair, and RNA processing.

Interestingly, one group of candidates which represses L1 includes the human silencing hub (HUSH) complex subunits and microrchidia 2 (MORC2). HUSH complex consists of M-phase phosphoprotein 8 (MPP8), Transgene Activation Suppressor (TASOR, also known as FAM208A), and Periphilin (PPHLN1). The HUSH complex recruits the H3K9me3 methyltransferase SETDB1 to repress genes, and the ATPase MORC2 to compact chromatin.[92,93] MORC2 is a member of the microrchidia (MORC) protein family involved in transposon silencing in plants and mice.[94,95]

Importantly, HUSH/MORC2 was found to bind to young, full-length L1 in transcriptionally permissive euchromatic regions, and promote H3K9me3 formation for silencing.[65] These young, full-length L1s frequently reside within introns of transcriptionally active genes. Surprisingly, these L1 methylations occur within introns of transcriptionally active genes and lead to modest, but significant downregulation of host gene expression in a HUSH/MORC2-dependent manner. This result has important implications for host genes, as this allows L1 to act indirectly as a transcription repressor that could influence the expression level of host genes whose intron it resides in. This silencing is in contrast to that which occurs within default heterochromatin, as this HUSH/MORC2-dependent silencing of young L1 occurs within introns of transcriptionally active genes within a permissive euchromatic region context. Very relevant to these ideas is that HUSH is well known to silence transgenes.[96]

Additional evidence further established the role of the HUSH complex in restricting L1. TASOR was found to bind to both endogenous retroviruses (ERVs) and L1s, and is required for H3K9me3 deposition over L1 and repetitive exons in transcribed genes.[97] This study also identified TASOR as a multifunctional protein related to PARP (polyA ribose polymerase) that directs HUSH assembly and epigenetic regulation of repetitive sequences in the genome. In a separate study involving mouse embryonic stem cells, TASOR was found to be required to repress young L1s (including those residing in introns of expressed host genes). TASOR works in concert with TRIM28, a well-known epigenetic repressor protein involved in transcriptional silencing of retroviruses, to repress young L1s.[98] One of the known functions of TRIM28 is to recruit SetDB1, the H3K9 methyltransferase. Genes repressed by both TRIM28 and TASOR are evolutionarily young and have tissue-specific expression. Thus, the HUSH complex and TRIM28 also work in euchromatic regions for epigenetic repression, and not in strict heterochromatin. This observation is in line with proposed functions of the HUSH complex (repression of L1s within introns of transcriptionally active host genes).[65]

The HUSH complex was found to control type I interferon (IFN) signaling in human cells[91] that is dependent on cytosolic nucleic acid-sensing proteins, such as MDA5, capable of detecting dsRNA. The authors suggested a role for L1 expression[24,99] in the upregulation of type I IFN signaling. The depletion of the HUSH complex subunit MPP8 results in overexpression of L1 and LTRs.[91] Interestingly, certain full-length hominid-specific L1s, such as L1Hs and L1PA2, produce L1 bidirectional RNA transcripts (sense and antisense RNAs) that are further enriched in MPP8-depleted cells. Perhaps these bidirectional L1 RNA transcripts form dsRNAs that are further recognized by the cytosolic nucleic acid-sensing component MDA5 (known to detect dsRNA more than 2 kb in length).[8] This result is surprising, because earlier work on primate-specific ORF0, which is encoded between nucleotides 452 and 236 of L1 5′ UTR in the antisense direction, had the potential to generate only much shorter antisense L1 transcripts of only a few hundred base pairs.[19] Alternative possible mechanisms are that HUSH recognizes cDNA or reflects L1 RT activity, since nucleoside RT inhibitors reduce the IFN response in all of these systems; or potentially, the signal event could even be downstream DNA damage resulting from retrotransposition. Using shRNAs against L1, the authors demonstrated that L1 expression is at least partially responsible for upregulation of IFN signaling when MPP8 is depleted. This further strengthened the connection between the HUSH/MORC2 silencing complex and the restriction of L1 activity. This work has implications for cancer cells that have lost epigenetic silencing with upregulated L1s. A recent study further implicated epigenetic silencing in the modulation of L1s in cancer cells and its relation to immunotherapy.[100] Tumors use different strategies to evade the immune system, and this study identified SETDB1 and members of the HUSH and KAP1 complexes as key modulators of immune escape in this cancer model. SETDB1 loss leads to TE derepression, suggesting that SETDB1 keeps the impact of TE in check in cells.

Host cells also employ transcriptional repression to restrict L1. The Krüppel-associated Box-containing (KRAB) Zinc-Finger Protein 93 (ZNF93) is known to bind within the 5' UTRs of L1PA3 and L1PA4 to repress expression of L1.[61] A slew of additional KRAB Zinc finger proteins are similarly implicated in L1 downregulation.[101] On an evolutionary timescale, this family of proteins can rapidly evolve to downregulate specific repeat sequence families, and has done so multiple times in the history of L1 evolution as well as control of endogenized retroviruses, and has its primary function in TE control, but has subsequently been exapted into developmental regulation.[102] However, KRAB zinc finger protein expression might be limited to particular periods of development, and its importance in alterations seen in cancer cells remains to be determined. Additional proteins, including transcription factors, that bind to L1, especially at the 5' UTR, have been identified using a bioinformatics approach termed "MapRRCon" (mapping repeat reads to a consensus) on the ENCODE ChIP-seq datasets.[103] Oncoprotein Myc and boundary/chromatin insulating factor CTCF were found to bind to the L1 DNAs in different cell types. Interestingly, RNA seq data showed that Myc RNA expression is inversely correlated to L1 RNA expression in breast and ovarian tumors. CTCF binding was seen at the 5' UTR and 3' UTR of L1s, and also colocalizes with Myc and RNA polymerase II.

P53 is one of the most studied transcription factors and tumor suppressors, and is mostly known in relation to carcinogenesis. Interestingly, multiple mechanisms have been proposed through which p53 may affect retrotransposition. That P53 restrains retro-transposon activity for genome stability has previously been demonstrated in *Drosophila*, *zebrafish*, and human cell lines.[104,105] An elevated amount of L1 ORF1p was also detected in a mouse model of p53-deficient, myc-driven liver cancer. In the *Drosophila* germline, transposon expression is observed in the p53⁻ germline.[104] Interestingly, such transposon RNA accumulation required functional meiotic recombination, as mutation of Spo11 (whose function is required to initiate meiotic recombination) abolished increased RNA expression for several transposons. Such an observation, together with the diverse but poorly understood roles of p53 prompted Wylie *et al.* to speculate on the primordial function of p53. Perhaps, the most primitive role of p53 is unrelated to its role in cancer, but in restricting retrotransposon mobility.[104,106] p53 has been shown to cooperate with the PIWI/piRNA pathway to restrain retrotransposon activity in the germline.[104] A recent study provides evidence to further strengthen the role played by p53 in repressing L1.[107] P53 binds directly to the L1 5′ UTR and facilitates establishment of repressive histone modifications in human cells.[107] Taken together, these findings suggest an important role for p53 in preventing L1 expression, which is compromised by p53 loss in carcinogenesis. The role of p53 in regulating L1 is also supported by a study which showed that, when L1 RNA is forced to express in non-transformed cells, the non-transformed cells undergo a p53-dependent growth arrest.[24] Interestingly, L1 expression in this system also activates interferon signaling, like viral infections, likely reinforcing limitations on the growth of L1(+) cells.

An exciting discovery about L1 restriction concerns the role of splicing in L1 RNA, and the discovery of spliced integrated retro-transposed elements (SpIREs) in the human genome.[108] Previous studies had documented the presence of splice donor (SD) and splice acceptor (SA) sites in human L1 RNA.[20,109,110] Larson *et al.*[108] performed *in vitro* studies based on potential spliced donor (SD) and splice acceptor (SA) sites within L1 5′UTR and ORF1, and also performed a human genome analysis to seek evidence for novel insertions of spliced integrated L1 sequences. These studies revealed that human L1 indeed contains functional splice donor (SD) and splice acceptor

(SA) sites within its 5' UTR and ORF1, although these are inefficiently employed.[108] These splice sites could generate spliced L1 mRNAs where splicing occurs completely within the 5' UTR or between 5' UTR and ORF1, generating spliced non-full-length L1 mRNAs. These spliced L1 mRNAs can retrotranspose and generate spliced integrated retrotransposed elements. Interestingly, SpIREs represent about 2% of annotated full-length primate-specific L1s in the human genome reference sequence. Both intra-5' UTR and 5' UTR/ORF1 SpIREs lack *cis*-acting transcription factor binding sites and thus have reduced promoter activity. The 5' UTR/ORF1 SpIREs produce non-functional ORF1p variants, indicating that these SpIREs cannot undergo new rounds of retrotransposition. Importantly, most L1 SpIREs contain ZNF93-binding site deletions, suggesting pressure to evade ZNF93 repression.[61] SpiREs provide a mechanism for the generation of truncated L1s in our genome, and illustrate splicing as a way for human genomes to restrict L1s.

6.3 *L1 ORF2 expression*

Even though L1 is upregulated in more than half of cancers, and ORF1 protein expression serves as a hallmark of cancer,[111] the expression of L1 ORF2p and its detection in L1 RNPs remain elusive.[90] This may relate to inefficient ORF2p production from the L1 transcript, although rapid clearance of ORF2p is also a possible contributor. ORF2p is translated through an unconventional mechanism involving ribosomal reinitiation on the bicistronic L1 RNA transcript.[112] Although several studies have tried to develop reagents to detect L1 ORF2p, endogenous ORF2p is still not detected by routine mass spectrometry methods in human cancers, even though ORF1p is readily detected by those methods.[90] Because ORF2p-associated DNA damage and RT activity limit cell growth,[24,90,113] limiting ORF2p expression may be an important dependency of cancer cells.

Using overexpression systems, L1 ORF2p can be detected, and imaging a tagged L1 ORF2p shows its presence in a rather low percentage of cells in a population.[47] There seems to be two distinct states of ORF2p: a cytoplasmic ORF2p population which probably represents the canonical L1 RNA/ORF1p/ORF2p RNPs and a nuclear ORF2p population which is observed in some cells.[66] Interestingly, the nuclear ORF2p staining is devoid of ORF1p, and this ORF2p-containing protein complex consists of at least four replication-associated proteins: PCNA, PURA/B, TOP1, and PARP1.[66] The presence of nuclear ORF2p requires ORF1, presumably for nuclear entry, and perhaps the ORF1p is replaced during reverse transcription with DNA replication and repair factors.

A separate study also detected ORF2p in both the cytoplasm and nucleus. ORF2p could be detected in the nucleus in a low percentage of cells, and is also devoid of ORF1p.[53] L1 mRNA, ORF1p, and ORF2p seem to enter the nucleus during mitosis or nuclear envelope breakdown, and can be detected in G1. L1 retrotransposition seems to peak during DNA replication in S phase.[46] RT and TPRT both require dNTPs, which are dependent on the ribonucleotide reductase enzyme (RNR). Both dNTPs and RNR are tightly regulated and are available during S phase. Mimosine, which inhibits RNR, also inhibits retrotransposition, supporting the idea that limited nucleotide pools can restrict L1 retrotransposition to S phase. L1 ORF2p binds replication fork components (PCNA and MCM proteins), further supporting the link between L1 retrotransposition and DNA replication forks.[46]

6.4 *DNA replication and DNA repair*

Proteomic studies have revealed multiple host factors that interact with L1 ORF1p, ORF2p, or L1 RNA[54,114,115] including RNA-binding proteins, RNA processing factors, and DNA replication factors. These include PABPC1 and MOV10, UPF1 (a key nonsense-mediated decay factor), and PCNA (the polymerase-delta-associated sliding DNA clamp).[54,66]

Recent progress has identified many retrotransposition[23,54,65-67] host factors involved in DNA repair and DNA replication (especially those related to replication forks and their repair). Factors involved in Nucleotide Excision Repair pathways, such as ERCC1-XPF, XPD, XPA, and the lesion-binding protein, XPC, have also been shown to play a role in limiting L1 retrotransposition, as *de novo* L1 inserts and their genomic locations in NER-deficient cells demonstrated the presence of abnormally large duplications at the site of insertion.[116] New functions for the NER pathway in the maintenance of genome integrity were proposed: limitation of insertional mutations caused by retrotransposons and the prevention of potentially mutagenic large genomic duplications at the sites of retrotransposon insertion events.

Interestingly, BRCA1 (breast cancer 1) seems to inhibit L1 ORF2 protein translation and levels in the cytoplasm, most likely through its binding of L1 mRNA.[23] This happens mainly in S/G2/M when BRCA1 is highly expressed. BRCA1 is known to exhibit translational control for other cytoplasmic proteins.[66] BRCA-1 siRNA treated cell lines have increased L1 ORF2p expression.[23] This upregulation might provide opportunities to further characterize the physical state of ORF2p in mammalian cells. L1 ORF2 protein level is inversely correlated to that of BRCA1: BRCA1 is most highly expressed during S/G2, when ORF2p is lowly expressed. This discovery might help explain why L1 is upregulated in cancers that have defective HR pathways; investigating expression in human breast and ovarian tumor specimens with BRCA1 and BRCA2 deficiencies would help shed light on these relationships.

Mita *et al.*[23] developed a microscopy-based retrotransposition assay in HeLa M2 cells, and identified DSB repair BRCA1 and Fanconi anemia (FA) factors as regulators of L1 activity. These two host factors are active in the S/G2 phase, and were found to act as potent inhibitors of retrotransposition at this phase. siRNA experiments showed that knockdown of homologous recombination (HR) factors, and to a lesser extent microhomology mediated end-joining factors, increases L1 retrotransposition frequency, suggesting that these pathways inhibit L1 retrotransposition. There may be competition at the replication fork between TPRT and HR factors including BRCA1 and FA. DNA repair proteins limit TPRT, and seem to protect replication forks from being used for L1 retrotransposition. This discovery has potential implications for upregulation of L1 in cancers with mis-regulated HR pathways (e.g., mutated *BRCA1* or *BRCA2*).

Similar results were reported in a related study: many DNA damage/repair factors, particularly the FA factors, repress L1 activity.[65]

Interestingly, in contrast, genes implicated in the non-homologous end joining (NHEJ) repair pathway seemingly promoted L1 retrotransposition, aligning with previous observations that mutations in some of the identified NHEJ factors were found to result in decreased retrotransposition frequencies.[65]

The conflict between DNA replication and L1 retrotransposition is further illustrated in Ardeljan *et al.* in their clear-cut finding that L1 expression induces a p53-dependent growth arrest.[24] CRISPR/CAS knockout screens were performed to reveal dependencies of p53-deficient, L1+ cells. These cells require replication-coupled DNA repair pathways (FA), replication stress signaling (ATRIP, 9-1-1 complex components), and replication fork restart factors (BLM and WRN helicases) to grow. Interestingly, these same sets of factors have also been recovered as those that are required for L1 retrotransposition.[23,65] Importantly, both L1's EN and RT activities are required to induce replication stress and to activate the FA pathway. The authors hypothesized that, based on what is known about TPRT, perhaps the branched L1 insertion intermediate structures create physical blockades to replication fork progression and promote fork stalling. In contrast to retrotransposition-based assays used by Liu *et al.*[65] and Mita *et al.*[23], done over a few days, the cell fitness screens done here took place over a period of weeks, during which time selection against cells undergoing retrotransposition presumably depleted their populations.

6.5 *L1 TPRT and integration*

De novo L1 insertion site sequences were analyzed using high throughput sequencing in HeLa cells[26] and several other cell lines including teratocarcinoma and embryonic stem cells.[25] In contrast to native L1 insertion site studies, these *de novo* L1 transposition event studies provide a unique opportunity to allow for the analysis of L1 insertion site preference without being influenced by evolutionary selection, and can be contrasted with samples taken from tumors, which may be more constrained.[117] Perhaps the most surprising result was the overall lack of chromatin or global sequence preferences for the insertion site. Instead, the target site selection seems to be primarily guided by the degenerate consensus sequence preferred by L1 endonuclease, 5′ TTTT/AA-3 ′, which produces a 3′ OH group that could be used for TPRT. The L1 insertion site has no preference for transcription or chromatin status of the insertion site. Instead, because of this simple preference by L1 EN, in combination with the existing nucleotide composition bias in the genome, this still results in preferred target integration sites for L1. As this preferred target sequence is found to be overrepresented in the DNA lagging strand, both studies were able to correlate L1 insertions with L1 replication fork directionality: L1 insertion is predominantly found in the lagging strand. Consistent with the DNA replication fork serving as a preferred target site for *de novo* L1 insertion is the analysis of L1 insertions in FANCD2 cell lines defective for DNA repair and replication fork stability. Interestingly, in this case, even an L1 EN mutant could still retrotranspose, using the free 3′ OH end of Okazaki fragments to initiate TPRT.[25] Taken together with studies that show that L1 retrotransposition occurs in S phase, and that L1 ORF2 interacts with PCNA, PARP1, and proteins at the DNA replication fork, these studies further strengthen the current model that L1 retrotransposition is tightly coupled to DNA replication. PARP1 is a sensor for unligated Okazaki fragments,[118] suggesting that L1 might exploit a stalled replication fork. That L1 retrotransposition occurs during S phase based on nucleotide sequence preference also helps explain the lack of chromatin state preference for L1 insertion site.

7. Conclusions

Progress in understanding host factor relationships with both LTR and non-LTR retrotransposons has elucidated a plethora of mechanisms employed by host cells to restrict retrotransposons. Though some host factors are required for retrotransposition, L1 retrotransposons are subjected to host factor restriction throughout the entire life cycle, starting from epigenetic silencing (DNA methylation and histone modifications) at the very early stage until the very final step of L1 retrotransposition, involving limits to insertions by DNA

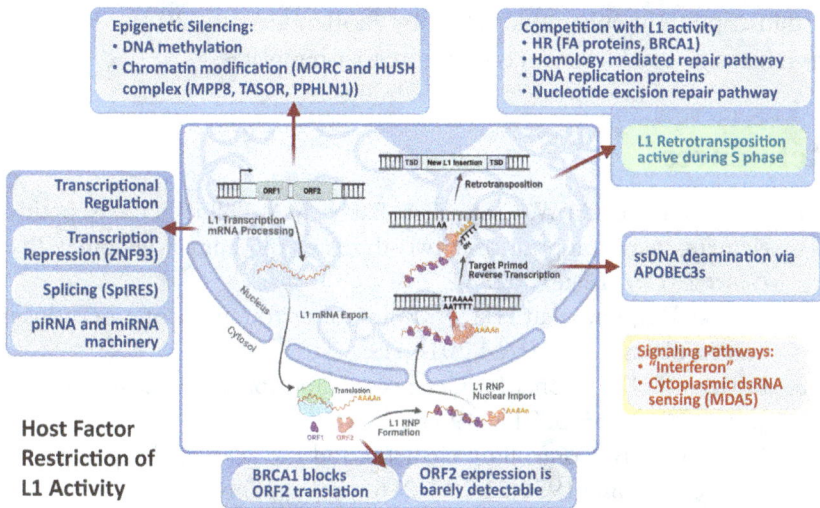

Figure 5. Host cell restriction of L1 activity occurs throughout the entire life cycle of L1 retrotransposon.

Host cells employ diverse mechanisms to restrict L1 activity during the entire life cycle of L1 retrotransposon. These mechanisms include the following: epigenetic silencing (via DNA methylation and chromatin modifications), transcriptional repression (including ZNF93 and p53), splicing, inhibition of L1 ORF2 translation by BRCA1, ssDNA deamination by ApoBEC3, DNA repair factors (including Fanconi Anemia proteins and BRCA1), and DNA replication factors (including PCNA). A non-homologous end joining pathway has a dual role, and could also stimulate L1 retrotransposition. PABPC1, the poly A-binding protein, could also stimulate L1 retrotransposition. Figure prepared in BioRender.

replication and repair factors (Fig. 5). As L1 activity is increasingly being implicated in cancers (see the chapter by Ardeljan and Burns) and other human diseases (see the chapter by Kazazian), both in somatic cells and the germline, progress in our understanding of host factor interaction with L1 and the mechanisms used by the host to restrict L1s may have important therapeutic implications.

Acknowledgments

KHB and JDB gratefully acknowledge the support of the NIH for their work on retrotransposons. The authors thank Meghan O'Keefe for her tireless assistance with editing. Figures 2, 3 and 5 are adapted from "Regulation of transcription in eukaryotic cells", by BioRender.com (2021). Retrieved from https://app.biorender.com/biorender-templates.

References

1. de Koning AP, Gu W, Castoe TA, Batzer MA, Pollock DD. Repetitive elements may comprise over two-thirds of the human genome. *PLoS Genet* 2011;7(12):e1002384.
2. Hancks DC, Kazazian HH, Jr. Roles for retrotransposon insertions in human disease. *Mob DNA* 2016;7:9.
3. Britten RJ, Davidson EH. Gene regulation for higher cells: a theory. *Science* 1969;165(3891):349–57.
4. McClintock B. Controlling elements and the gene. *Cold Spring Harb Symp Quant Biol* 1956;21:197–216.
5. Bourque G. Transposable elements in gene regulation and in the evolution of vertebrate genomes. *Curr Opin Genet Dev* 2009;19(6): 607–12.
6. Chuong EB, Elde NC, Feschotte C. Regulatory activities of transposable elements: from conflicts to benefits. *Nat Rev Genet* 2017; 18(2):71–86.
7. Boeke JD, Stoye JP. Retrotransposons, endogenous retroviruses, and the evolution of retroelements. In: Coffin JM, Hughes SH, Varmus HE, eds. Retroviruses. NY: Cold Spring Harbor; 1997.
8. Judd J, Sanderson H, Feschotte C. Evolution of mouse circadian enhancers from transposable elements. *bioRxiv* 2020:2020.2011. 2009.375469.
9. Mu X, Ahmad S, Hur S. Endogenous retroelements and the host innate immune sensors. *Adv Immunol* 2016;132:47–69.
10. Jung YD, Ahn K, Kim YJ, Bae JH, Lee JR, Kim HS. Retroelements: molecular features and implications for disease. *Genes Genet Syst* 2013;88(1):31–43.
11. Grau JH, Poustka AJ, Meixner M, Plotner J. LTR retroelements are intrinsic components of transcriptional networks in frogs. *BMC Genomics* 2014;15:626.
12. Coates BS, Fraser LM, French B, Sappington TW. Proliferation and copy number variation of BEL-like long terminal repeat retrotransposons within the Diabrotica virgifera virgifera genome. *Gene* 2014;534(2):362–70.

13. de la Chaux N, Wagner A. BEL/Pao retrotransposons in metazoan genomes. *BMC Evol Biol* 2011;11:154.

14. Rodriguez F, Kenefick AW, Arkhipova IR. LTR-retrotransposons from bdelloid rotifers capture additional ORFs shared between highly diverse retroelement Types. *Viruses* 2017;9(4).

15. Czaja W, Bensasson D, Ahn HW, Garfinkel DJ, Bergman CM. Evolution of Ty1 copy number control in yeast by horizontal transfer and recombination. *PLoS Genet* 2020;16(2):e1008632.

16. Neuveglise C, Feldmann H, Bon E, Gaillardin C, Casaregola S. Genomic evolution of the long terminal repeat retrotransposons in hemiascomycetous yeasts. *Genome Res* 2002;12(6):930–43.

17. Liti G, Peruffo A, James SA, Roberts IN, Louis EJ. Inferences of evolutionary relationships from a population survey of LTR-retrotransposons and telomeric-associated sequences in the Saccharomyces sensu stricto complex. *Yeast* 2005;22(3):177–92.

18. Bleykasten-Grosshans C, Friedrich A, Schacherer J. Genome-wide analysis of intraspecific transposon diversity in yeast. *BMC Genomics* 2013;14:399.

19. Denli AM, Narvaiza I, Kerman BE, et al. Primate-specific ORF0 contributes to retrotransposon-mediated diversity. *Cell* 2015;163(3): 583–93.

20. Belancio VP, Hedges DJ, Deininger P. LINE-1 RNA splicing and influences on mammalian gene expression. *Nucleic Acids Res* 2006;34(5):1512–21.

21. Akoh CC, Lee GC, Liaw YC, Huang TH, Shaw JF. GDSL family of serine esterases/lipases. *Prog Lipid Res* 2004;43(6):534–52.

22. Schneider AM, Schmidt S, Jonas S, Vollmer B, Khazina E, Weichenrieder O. Structure and properties of the esterase from non-LTR retrotransposons suggest a role for lipids in retrotransposition. *Nucleic Acids Res* 2013;41(22):10563–72.

23. Mita P, Sun X, Fenyo D, et al. BRCA1 and S phase DNA repair pathways restrict LINE-1 retrotransposition in human cells. *Nat Struct Mol Biol* 2020;27(2):179–91.

24. Ardeljan D, Steranka JP, Liu C, et al. Cell fitness screens reveal a conflict between LINE-1 retrotransposition and DNA replication. *Nat Struct Mol Biol* 2020;27(2):168–78.

25. Flasch DA, Macia A, Sanchez L, et al. Genome-wide de novo L1 retrotransposition connects endonuclease activity with replication. *Cell* 2019;177(4):837–51 e828.

26. Sultana T, van Essen D, Siol O, *et al.* The landscape of L1 retrotransposons in the human genome is shaped by pre-insertion sequence biases and post-insertion selection. *Mol Cell.* 2019;74(3):555–70 e557.

27. Jiang J, Zhao L, Yan L, *et al.* Structural features and mechanism of translocation of non-LTR retrotransposons in Candida albicans. *Virulence* 2014;5(2):245–52.

28. Dong C, Poulter RT, Han JS. LINE-like retrotransposition in Saccharomyces cerevisiae. *Genetics* 2009;181(1):301–11.

29. Fujiwara H. Site-specific non-LTR retrotransposons. *Microbiol Spectr* 2015;3(2):MDNA3-0001-2014.

30. Atkins JF, Loughran G, Bhatt PR, Firth AE, Baranov PV. Ribosomal frameshifting and transcriptional slippage: from genetic steganography and cryptography to adventitious use. *Nucleic Acids Res* 2016;44(15): 7007–78.

31. Wilhelm FX, Wilhelm M, Gabriel A. Reverse transcriptase and integrase of the Saccharomyces cerevisiae Ty1 element. *Cytogenet Genome Res* 2005;110(1–4):269–87.

32. Hughes SH. Reverse transcription of retroviruses and LTR retrotransposons. *Microbiol Spectr* 2015;3(2):MDNA3-0027-2014.

33. Mules EH, Uzun O, Gabriel A. In vivo Ty1 reverse transcription can generate replication intermediates with untidy ends. *J Virol* 1998; 72(8):6490–503.

34. Beck CR, Collier P, Macfarlane C, *et al.* LINE-1 retrotransposition activity in human genomes. *Cell* 2010;141(7):1159–70.

35. Gabriel A, Willems M, Mules EH, Boeke JD. Replication infidelity during a single cycle of Ty1 retrotransposition. *Proc Natl Acad Sci USA* 1996;93(15):7767–71.

36. Feng YX, Moore SP, Garfinkel DJ, Rein A. The genomic RNA in Ty1 virus-like particles is dimeric. *J Virol* 2000;74(22):10819–21.

37. Checkley MA, Mitchell JA, Eizenstat LD, Lockett SJ, Garfinkel DJ. Ty1 gag enhances the stability and nuclear export of Ty1 mRNA. *Traffic* 2013;14(1):57–69.

38. Chapman KB, Bystrom AS, Boeke JD. Initiator methionine tRNA is essential for Ty1 transposition in yeast. *Proc Natl Acad Sci USA* 1992;89(8):3236–40.

39. Xu H, Boeke JD. Localization of sequences required in cis for yeast Ty1 element transposition near the long terminal repeats: analysis of mini-Ty1 elements. *Mol Cell Biol* 1990;10(6):2695–2702.

40. Bolton EC, Coombes C, Eby Y, Cardell M, Boeke JD. Identification and characterization of critical cis-acting sequences within the yeast Ty1 retrotransposon. *RNA* 2005;11(3):308–22.
41. Huang Q, Purzycka KJ, Lusvarghi S, Li D, Legrice SF, Boeke JD. Retrotransposon Ty1 RNA contains a 5'-terminal long-range pseudoknot required for efficient reverse transcription. *RNA* 2013; 19(3):320–32.
42. Gumna J, Purzycka KJ, Ahn HW, Garfinkel DJ, Pachulska-Wieczorek K. Retroviral-like determinants and functions required for dimerization of Ty1 retrotransposon RNA. *RNA Biol* 2019;16(12):1749–63.
43. Cristofari G, Bampi C, Wilhelm M, Wilhelm FX, Darlix JL. A 5'-3' long-range interaction in Ty1 RNA controls its reverse transcription and retrotransposition. *EMBO J* 2002;21(16):4368–79.
44. Matsuda E, Garfinkel DJ. Posttranslational interference of Ty1 retrotransposition by antisense RNAs. *Proc Natl Acad Sci USA* 2009; 106(37):15657–62.
45. Andrzejewska A, Zawadzka M, Gumna J, Garfinkel DJ, Pachulska-Wieczorek K. In vivo structure of the Ty1 retrotransposon RNA genome. *Nucleic Acids Res* 2021;49(5):2878–93.
46. Gamache ER, Doh JH, Ritz J, *et al.* Structure-Function model for kissing loop interactions that initiate dimerization of Ty1 RNA. *Viruses* 2017;9(5).
47. Sandmeyer S, Patterson K, Bilanchone V. Ty3, a position-specific retrotransposon in budding yeast. *Microbiol Spectr* 2015;3(2): MDNA3-0057-2014.
48. Bonnet A, Lesage P. Light and shadow on the mechanisms of integration site selection in yeast Ty retrotransposon families. *Curr Genet* 2021.
49. Sultana T, Zamborlini A, Cristofari G, Lesage P. Integration site selection by retroviruses and transposable elements in eukaryotes. *Nat Rev Genet* 2017;18(5):292–308.
50. Craigie R, Bushman FD. Host factors in retroviral integration and the selection of integration target sites. *Microbiol Spectr* 2014;2(6).
51. Khazina E, Truffault V, Buttner R, Schmidt S, Coles M, Weichenrieder O. Trimeric structure and flexibility of the L1ORF1 protein in human L1 retrotransposition. *Nat Struct Mol Biol* 2011;18(9):1006–14.
52. Khazina E, Weichenrieder O. Human LINE-1 retrotransposition requires a metastable coiled coil and a positively charged N-terminus in L1ORF1p. *Elife* 2018;7.

53. Mita P, Wudzinska A, Sun X, *et al.* LINE-1 protein localization and functional dynamics during the cell cycle. *Elife* 2018;7.

54. Taylor MS, LaCava J, Mita P, *et al.* Affinity proteomics reveals human host factors implicated in discrete stages of LINE-1 retrotransposition. *Cell* 2013;155(5):1034–48.

55. Mathias SL, Scott AF, Kazazian HH, Jr., Boeke JD, Gabriel A. Reverse transcriptase encoded by a human transposable element. *Science* 1991;254(5039):1808–10.

56. Feng Q, Moran JV, Kazazian HH, Jr., Boeke JD. Human L1 retrotransposon encodes a conserved endonuclease required for retrotransposition. *Cell* 1996;87(5):905–16.

57. Hoffman H, Skordalakes E. Crystallographic studies of telomerase. *Methods Enzymol* 2016;573:403–19.

58. Yoder JA, Walsh CP, Bestor TH. Cytosine methylation and the ecology of intragenomic parasites. *Trends Genet* 1997;13(8):335–40.

59. Jordan IK, Matyunina LV, McDonald JF. Evidence for the recent horizontal transfer of long terminal repeat retrotransposon. *Proc Natl Acad Sci U S A* 1999;96(22):12621–5.

60. Aravin AA, Hannon GJ, Brennecke J. The Piwi-piRNA pathway provides an adaptive defense in the transposon arms race. *Science* 2007;318(5851):761–4.

61. Jacobs FM, Greenberg D, Nguyen N, *et al.* An evolutionary arms race between KRAB zinc-finger genes ZNF91/93 and SVA/L1 retrotransposons. *Nature* 2014;516(7530):242–5.

62. Van Valen L. "A new evolutionary law" (PDF). *Evol Theory* 1973;1: 1–30.

63. Carroll L. *Through the Looking Glass.* London: MacMillan; 1872.

64. Koonin EV, Aravind L. Comparative genomics, evolution and origins of the nuclear envelope and nuclear pore complex. *Cell Cycle* 2009;8(13):1984–5.

65. Liu N, Lee CH, Swigut T, *et al.* Selective silencing of euchromatic L1s revealed by genome-wide screens for L1 regulators. *Nature* 2018; 553(7687):228–232.

66. Taylor MS, Altukhov I, Molloy KR, *et al.* Dissection of affinity captured LINE-1 macromolecular complexes. *Elife* 2018;7.

67. Briggs EM, McKerrow W, Mita P, Boeke JD, Logan SK, Fenyo D. RIP-seq reveals LINE-1 ORF1p association with p-body enriched mRNAs. *Mob DNA* 2021;12(1):5.

68. Levin HL. A novel mechanism of self-primed reverse transcription defines a new family of retroelements. *Mol Cell Biol* 1995;15(6):3310–17.
69. Cullen H, Schorn AJ. Endogenous retroviruses walk a fine line between priming and silencing. *Viruses* 2020;12(8).
70. Boeke JD, Devine SE. Yeast retrotransposons: finding a nice quiet neighborhood. *Cell* 1998;93(7):1087–9.
71. Mularoni L, Zhou Y, Bowen T, Gangadharan S, Wheelan SJ, Boeke JD. Retrotransposon Ty1 integration targets specifically positioned asymmetric nucleosomal DNA segments in tRNA hotspots. *Genome Res* 2012;22(4):693–703.
72. Baller JA, Gao J, Stamenova R, Curcio MJ, Voytas DF. A nucleosomal surface defines an integration hotspot for the Saccharomyces cerevisiae Ty1 retrotransposon. *Genome Res* 2012;22(4):704–13.
73. Zou S, Ke N, Kim JM, Voytas DF. The Saccharomyces retrotransposon Ty5 integrates preferentially into regions of silent chromatin at the telomeres and mating loci. *Genes Dev* 1996;10(5):634–45.
74. Zhu Y, Dai J, Fuerst PG, Voytas DF. Controlling integration specificity of a yeast retrotransposon. *Proc Natl Acad Sci U S A* 2003;100(10): 5891–5.
75. Asif-Laidin A, Conesa C, Bonnet A, et al. A small targeting domain in Ty1 integrase is sufficient to direct retrotransposon integration upstream of tRNA genes. *EMBO J* 2020;39(17):e104337.
76. Kenna MA, Brachmann CB, Devine SE, Boeke JD. Invading the yeast nucleus: a nuclear localization signal at the C terminus of Ty1 integrase is required for transposition in vivo. *Mol Cell Biol* 1998;18(2):1115–24.
77. Griffith JL, Coleman LE, Raymond AS, et al. Functional genomics reveals relationships between the retrovirus-like Ty1 element and its host Saccharomyces cerevisiae. *Genetics* 2003;164(3):867–79.
78. Rowley PA, Patterson K, Sandmeyer SB, Sawyer SL. Control of yeast retrotransposons mediated through nucleoporin evolution. *PLoS Genet* 2018;14(4):e1007325.
79. Manhas S, Ma L, Measday V. The yeast Ty1 retrotransposon requires components of the nuclear pore complex for transcription and genomic integration. *Nucleic Acids Res* 2018;46(7):3552–78.
80. Salinero AC, Knoll ER, Zhu ZI, Landsman D, Curcio MJ, Morse RH. The Mediator co-activator complex regulates Ty1 retromobility by controlling the balance between Ty1i and Ty1 promoters. *PLoS Genet* 2018;14(2):e1007232.

81. Ahn HW, Tucker JM, Arribere JA, Garfinkel DJ. Ribosome biogenesis modulates Ty1 copy number control in Saccharomyces cerevisiae. *Genetics* 2017;207(4):1441–56.

82. Doh JH, Lutz S, Curcio MJ. Co-translational localization of an LTR-retrotransposon RNA to the endoplasmic reticulum nucleates virus-like particle assembly sites. *PLoS Genet* 2014;10(3):e1004219.

83. Miki Y, Nishisho I, Horii A, *et al.* Disruption of the APC gene by a retrotransposal insertion of L1 sequence in a colon cancer. *Cancer Res* 1992;52(3):643–5.

84. Kazazian HH, Jr. Mobile DNA transposition in somatic cells. *BMC Biol* 2011;9:62.

85. Sassaman DM, Dombroski BA, Moran JV, *et al.* Many human L1 elements are capable of retrotransposition. *Nat Genet* 1997;16(1):37–43.

86. Brouha B, Schustak J, Badge RM, *et al.* Hot L1s account for the bulk of retrotransposition in the human population. *Proc Natl Acad Sci USA* 2003;100(9):5280–5285.

87. Uriu K, Kosugi Y, Suzuki N, Ito J, Sato K. Elucidation of the complicated scenario of primate APOBEC3 gene evolution. *J Virol* 2021.

88. Molaro A, Malik HS, Bourc'his D. Dynamic evolution of de novo DNA methyltransferases in rodent and primate genomes. *Mol Biol Evol* 2020;37(7):1882–92.

89. Bruno M, Mahgoub M, Macfarlan TS. The arms race between KRAB-zinc finger proteins and endogenous retroelements and its impact on mammals. *Annu Rev Genet* 2019;53:393–416.

90. Ardeljan D, Wang X, Oghbaie M, *et al.* LINE-1 ORF2p expression is nearly imperceptible in human cancers. *Mob DNA* 2020;11:1.

91. Tunbak H, Enriquez-Gasca R, Tie CHC, *et al.* The HUSH complex is a gatekeeper of type I interferon through epigenetic regulation of LINE-1s. *Nat Commun* 2020;11(1):5387.

92. Douse CH, Bloor S, Liu Y, *et al.* Neuropathic MORC2 mutations perturb GHKL ATPase dimerization dynamics and epigenetic silencing by multiple structural mechanisms. *Nat Commun* 2018;9(1):651.

93. Tchasovnikarova IA, Timms RT, Douse CH, *et al.* Hyperactivation of HUSH complex function by Charcot-Marie-Tooth disease mutation in MORC2. *Nat Genet* 2017;49(7):1035–44.

94. Moissiard G, Cokus SJ, Cary J, *et al.* MORC family ATPases required for heterochromatin condensation and gene silencing. *Science* 2012;336(6087):1448–1451.

95. Pastor WA, Stroud H, Nee K, *et al.* MORC1 represses transposable elements in the mouse male germline. *Nat Commun* 2014;5:5795.
96. Tchasovnikarova IA, Timms RT, Matheson NJ, *et al.* GENE SILENCING. Epigenetic silencing by the HUSH complex mediates position-effect variegation in human cells. *Science* 2015; 348(6242):1481–1485.
97. Douse CH, Tchasovnikarova IA, Timms RT, *et al.* TASOR is a pseudo-PARP that directs HUSH complex assembly and epigenetic transposon control. *Nat Commun* 2020;11(1):4940.
98. Robbez-Masson L, Tie CHC, Conde L, *et al.* The HUSH complex cooperates with TRIM28 to repress young retrotransposons and new genes. *Genome Res* 2018;28(6):836–45.
99. De Cecco M, Ito T, Petrashen AP, *et al.* L1 drives IFN in senescent cells and promotes age-associated inflammation. *Nature* 2019;566(7742):73–8.
100. Griffin GK, Wu J, Iracheta-Vellve A, *et al.* Epigenetic silencing by SETDB1 suppresses tumour intrinsic immunogenicity. *Nature* 2021.
101. Imbeault M, Helleboid PY, Trono D. KRAB zinc-finger proteins contribute to the evolution of gene regulatory networks. *Nature* 2017;543(7646):550–4.
102. Ecco G, Imbeault M, Trono D. KRAB zinc finger proteins. *Development* 2017;144(15):2719–29.
103. Sun X, Wang X, Tang Z, *et al.* Transcription factor profiling reveals molecular choreography and key regulators of human retrotransposon expression. *Proc Natl Acad Sci U S A* 2018;115(24):E5526–35.
104. Wylie A, Jones AE, D'Brot A, *et al.* p53 genes function to restrain mobile elements. *Genes Dev* 2016;30(1):64–77.
105. Harris CR, Dewan A, Zupnick A, *et al.* p53 responsive elements in human retrotransposons. *Oncogene* 2009;28(44):3857–65.
106. Wylie A, Jones AE, Abrams JM. p53 in the game of transposons. *Bioessays* 2016;38(11):1111–6.
107. Tiwari B, Jones AE, Caillet CJ, Das S, Royer SK, Abrams JM. p53 directly represses human LINE1 transposons. *Genes Dev* 2020.
108. Larson PA, Moldovan JB, Jasti N, Kidd JM, Beck CR, Moran JV. Spliced integrated retrotransposed element (SpIRE) formation in the human genome. *PLoS Biol* 2018;16(3):e2003067.
109. Belancio VP, Roy-Engel AM, Deininger P. The impact of multiple splice sites in human L1 elements. *Gene* 2008;411(1–2):38–45.

110. Han JS, Szak ST, Boeke JD. Transcriptional disruption by the L1 retrotransposon and implications for mammalian transcriptomes. *Nature* 2004;429(6989):268–74.

111. Rodic N, Sharma R, Sharma R, *et al.* Long interspersed element-1 protein expression is a hallmark of many human cancers. *Am J Pathol* 2014;184(5):1280–1286.

112. Alisch RS, Garcia-Perez JL, Muotri AR, Gage FH, Moran JV. Unconventional translation of mammalian LINE-1 retrotransposons. *Genes Dev* 2006;20(2):210–24.

113. Gasior SL, Wakeman TP, Xu B, Deininger PL. The human LINE-1 retrotransposon creates DNA double-strand breaks. *J Mol Biol* 2006;357(5):1383–93.

114. Goodier JL, Cheung LE, Kazazian HH, Jr. Mapping the LINE1 ORF1 protein interactome reveals associated inhibitors of human retrotransposition. *Nucleic Acids Res* 2013;41(15):7401–19.

115. Moldovan JB, Moran JV. The zinc-finger antiviral protein ZAP inhibits LINE and Alu retrotransposition. *PLoS Genet* 2015;11(5):e1005121.

116. Servant G, Streva VA, Derbes RS, *et al.* The nucleotide excision repair pathway limits L1 retrotransposition. *Genetics* 2017;205(1):139–53.

117. Rodriguez-Martin B, Alvarez EG, Baez-Ortega A, *et al.* Pan-cancer analysis of whole genomes identifies driver rearrangements promoted by LINE-1 retrotransposition. *Nat Genet* 2020;52(3):306–319.

118. Hanzlikova H, Kalasova I, Demin AA, Pennicott LE, Cihlarova Z, Caldecott KW. The importance of poly(ADP-Ribose) polymerase as a sensor of unligated Okazaki fragments during DNA replication. *Mol Cell* 2018;71(2):319–31 e313.

119. Qi X, Sandmeyer S. Nonhomologous recombination: retrotransposons. In: Lennarz WJ, Lane MD, eds. Encyclopedia of Biological Chemistry (Second Edition). Waltham: Academic Press; 2012:283–91.

120. Peaston AE. Retrotransposons of vertebrates. In: Mahy BWJ, Van Regenmortel MHV, eds. Encyclopedia of Virology (Third Edition). Oxford: Academic Press; 2008:436–45.

Chapter 8

Retrotransposons in the Mammalian Brain

Tracy A. Bedrosian[†,*], Sara B. Linker[‡,*], and Fred H. Gage[‡]

1. Introduction

Retrotransposons are mobile genetic elements that are expressed and active in a variety of tissue types such as the liver and brain. For a detailed review on the fundamentals of retrotransposon biology.[1] In this chapter, we will focus on retrotransposons in the context of the mammalian brain, including their evolution, somatic retrotransposition, impact on disease, and cytotoxicity.

2. Exaptation of Retrotransposons for Functional Roles in the Brain

Exaptation of retrotransposons refers to the process by which retrotransposons are repeatedly co-opted by the host genome to serve a functional role. Below, we broadly describe a few of these scenarios in endogenous retroviruses (ERVs), Short Interspersed Nuclear Elements

---

[*]These authors contributed equally
[†]Institute for Genomic Medicine, Nationwide Children's Hospital, Department of Pediatrics, The Ohio State University, Columbus, OH, USA.
[‡]Laboratory of Genetics, The Salk Institute for Biological Studies, La Jolla, CA, USA.

(SINEs), and Long Interspersed Nuclear Elements (LINEs) in the context of the brain.

2.1 *ERVs and LTR-containing retrotransposons*

ERVs encode *gag*, *pol*, and *env* proteins and descend from exogenous retroviruses that invaded the host genome. The *env* gene is associated with infection in the vertebrate germline, while loss of *env* is associated with retrotransposition intracellularly. However, they are no longer active in the human genome.[2] Sections of these elements have repeatedly evolved functions in the host across multiple systems and organisms. For example, *Arc* is an immediate early gene that is important in synaptic plasticity and, subsequently, memory formation.[3] Mammalian *Arc* and homologs in other genera such as *Drosophila* (*dArc*) independently evolved from the Ty3/gypsy family of LTR-retrotransposons and contain *gag*-like coding regions.[4] *Arc* creates a capsid-like structure that shuttles RNA molecules across the synapse to connected neurons, aiding in synaptic plasticity.[4,5]

The convergent evolution of *Arc* underscores the utility of retrotransposon-encoded protein domains for host function. While it is unclear how many instances of exaptation have occurred in the neuronal context, Ty3/gypsy elements have generated other brain-expressed genes, hinting at a potential for more such instances to be discovered in the future. For example, *Sirh11/Zcchc16*, derived from the Ty3/gypsy *gag* gene, impacts noradrenergic function in the prefrontal cortex and impulsivity in knockout mice.[6] Two Mar family genes derived from Ty3/gypsy, *PEG10*, from a *gag* and *pol* fusion, and *LDOC1*, from the *gag* gene,[7] are highly expressed in the brain.[8,9] *PEG10* is expressed in a region-specific manner throughout the brain, including areas like the dorsal and median raphe, the locus coeruleus, and throughout the hypothalamus.[8] Experiments in non-neuronal cells indicate that *PEG10* interacts with the TGF-α superfamily of receptors, indicating a potential role in innate immunity.[10] *LDOC1* also interacts with innate immunity by repressing NF-κB.[11] However, due to the embryonic lethality of *PEG10* knockouts and limited experimentation on both *PEG10* and *LDOC1*, there is currently little to no information on the functions of these proteins in the brain.

ERVs have also been utilized by the host genome for transcriptional regulation. TRIM28, a transcriptional repressor, binds to primate-specific ERVs throughout the genome in neural progenitor cells (NPCs), directing genome-wide repression of neighboring genes.[12] ERVs themselves are also expressed within neuronal tissue and exhibit disease-associated transcriptional signatures. Human ERVs are enriched for Schizophrenia risk loci and are differentially expressed in a disease-dependent manner in patients diagnosed with schizophrenia, ADHD, and autism.[13–15] Furthermore, prenatal exposure to valproic acid, a drug that can induce autism-like features in mice when exposed prenatally, induces heightened expression of ERVs.[16] While it remains unclear what downstream impact aberrant ERV expression has on neuronal tissue, given the links with *cis* transcriptional regulation and innate immunity, these roles may be potential targets for future research.

2.2 *SINEs*

SINEs are a broad class of non-autonomous retrotransposons that have repeatedly been co-opted for function in the host genome. SINE families such as RSINEs, B1 and B2 in rodents, and *Alu* in humans are sensitive to cellular stress and are upregulated by events such as DNA damage by etoposide, heat shock, and neuronal activity,[17,18] and they can act as *cis* regulators of neuronal activity-dependent gene expression.[19,20]

While it is still unclear what role the heightened level of B2 or *Alu* has on neurons, there is considerable evidence for neuronal function of two intron-containing non-coding RNAs derived from B2 and *Alu*. These loci, termed brain cytoplasmic 1 (Bc1) and brain cytoplasmic 200 (Bc200; aka BCYRN1) in rodent and humans, respectively, are transcribed in neurons throughout the brain and have the unique feature of being trafficked down dendrites to synapto-dendritic domains.[21] In synaptic areas, they retain a function of the parent SINE element[22] to inhibit protein translation, and they do this in a neuronal activity-dependent manner.[23] BC1 knockout mice exhibit a range of neuronal deficits such as altered glutamatergic transmission, experience-dependent changes in synapses, electrophysiology, and behavior as indicated by excessive self-grooming, and deficits in discrimination and conflict learning paradigms.[24,25] Both mouse Bc1 and human BC200 directly associate with the fragile X syndrome protein (FMRP), a gene responsible for monogenic forms of Autism Spectrum Disorder (ASD). FMRP functions in part by inhibiting dendritic protein translation. Binding of FMRP is targeted to specific mRNAs such as *Map1b*, *Arc*, and *Camk2a*, by Bc1 through small homology regions between the SINE element and the corresponding mRNA.[26] This body of work shows that SINE elements have been separately co-opted by both primates and rodents to play a role in dendritically targeted protein translation with downstream impacts on neuronal function and animal behavior. Although much of the work in this area has focused on these two loci, there is evidence that other SINE elements from the same subfamilies are transcriptionally elevated in response to neuronal activity, although the downstream impact of this expression is still unknown.[17,27]

2.3 *LINEs*

LINEs are a class of autonomous retroelements that do not contain LTRs and are responsible for the majority of retrotransposition in humans including retrotransposition of non-autonomous SINEs. The active LINE family in humans and mouse is LINE-1 (L1). Discovering the functional importance of L1 in the mammalian brain is a burgeoning area of research. Both human and mouse NPCs support L1 retrotransposition, generating somatic genetic diversity.

3. Somatic Retrotransposition

Traditionally it was believed that every cell in an individual contained identical DNA sequences. Recently, however, it has become clear that each cell's genome can vary because of somatic mutations that arise throughout life, a phenomenon known as somatic mosaicism.[28] Depending on when a somatic mutation occurs, it may affect many or few cells. For example, a somatic mutation that occurs in an early progenitor cell during embryonic development could affect a large proportion of cells in the body. Alternatively, a mutation that occurs in a neural progenitor in a neurogenic niche of the adult brain may only affect a small handful of cells. Depending on the genomic location and type of the mutation, it can have a range of effects on cellular function by altering gene expression, transcript splicing, epigenetic modifications, or by generating novel protein content.[29] Particularly in a highly networked organ like the brain, a small number of somatic mutations affecting cell function could have far-reaching effects on neuronal circuitry. Somatic mosaicism is a feature of healthy brains, where it may contribute to normal functional diversity, but it is also a potential driver of disease when somatic mutations accumulate at high frequency.

Retrotransposons are one of several contributors to somatic mosaicism. In some contexts, somatic retrotransposon insertions occur at higher frequency than germline insertions; for example, the rate of Alu or L1 retrotransposition in the human germline is estimated at 1 insertion per 20–200 births, whereas the rate per somatic cell may be significantly higher depending on the tissue.[30] The brain, in particular, is a rich source of retrotransposon-derived somatic mosaicism, with at least one somatic retrotransposon insertion, or retrotransposon-associated mutation, for every two cells.[31] New retrotransposon insertions influence the transcriptome in a variety of ways (reviewed[29]), most notably by altering expression of mRNA and non-coding RNA (ncRNA). For example, L1 elements inserted into a gene in the sense orientation tend to decrease transcript abundance.[29] Also, transcription of L1 elements can create ncRNA, antisense mRNA, or double-stranded RNA, which can then influence gene expression.[29] A large proportion of exons containing an *Alu* insertion are alternatively spliced.[29] Besides classical retrotransposition, retrotransposons can also mediate somatic deletions through homologous recombination[29] and generate novel fusion proteins by making use of an antisense promoter, as in the case of L1.[32] Depending on the timing and location of a particular retrotransposon event, the consequences for a given brain cell or circuit may be varied.

3.1 *Development*

During development, neural stem cells rapidly divide to give rise to nearly 100 billion neurons in the mature brain. As these cells differentiate, expression of Sox2 declines and derepresses L1 transcription, providing an opportunity for retrotransposition to occur.[33] Cell division is an important permissive factor in contributing to retrotransposition because the nuclear envelope breakdown that occurs can allow an opportunity for L1 ribonucleoprotein complexes to be imported from the cytoplasm, where they can directly interact with genomic DNA.[34] Compared to cells from other organs and tissues, NPCs appear to support higher levels of retrotransposition, making brain development a key time for retrotransposon-mediated somatic mosaicism to arise.[35]

The probability of retrotransposition in the brain likely peaks during prenatal development because of the large number of rapidly dividing cells. This period is a critical window for modulation by endogenous and exogenous factors. For example, viral infections tend to increase expression of some retrotransposons, raising the question of whether prenatal infection modulates retrotransposition.[36] When pregnant mice are injected with Poly I:C, a compound that mimics viral double-stranded RNA to induce an immune response, their offspring end up with significantly more copies of L1 retrotransposons in the brain.[37] The consequences of this increase in L1 content are unclear, but prenatal viral infection also causes behavioral symptoms in mice reminiscent of schizophrenia.

A causative link between retrotransposon activity and behavior in mammals has yet to be established but is an active area of investigation. Other examples of environmental factors modulating retrotransposition can be drawn from studies in various organisms and cell culture. For example, studies suggest heat shock, exposure to toxic substances, drugs of abuse, and certain hormones may be able to modulate activity of retrotransposons (reviewed[38]).

By early postnatal development in mice and humans, most neurons that will compose the mature brain have already been formed; however, neurogenesis in the hippocampus and cerebellum is prolonged into the early postnatal weeks, providing another window for retrotransposons to mobilize and respond to modulating factors in the environment. In laboratory mice, there are natural variations in maternal behavior that influence the offspring in terms of gene expression and chromatin accessibility, brain plasticity, and adult behavioral phenotypes.[39] When pups are reared by a dam on the low end of the maternal spectrum, meaning one who spends more time away from the nest and less time caring for the pups, there is a decrease in methylation of L1 elements and a corresponding increase in L1 mRNA expression and copy number within the hippocampus, but not in other brain regions such as the frontal cortex, where neuronal division is expected to be complete.[40] Other retrotransposons, including SINE B1 and B2 elements, do not show such an increase. This finding suggests that neurons dividing during the early postnatal window are susceptible to acquiring somatic L1 retrotransposon insertions to a degree dependent on modulating factors. In humans, methylation studies support a similar notion, where children with a history of early life stress or trauma have reduced methylation of LINE and SINE retrotransposon loci.[41,42] Recently, a randomized study revealed that L1 is hypomethylated in preterm infants but that multi-sensory intervention, including infant massage and visual interaction, could restore L1 methylation levels.[43] Whether changes in methylation translate to changes in retrotransposition in humans is unknown, but these findings raise the possibility that somatic retrotransposition is a dynamic process that is responsive to environmental cues.

3.2 *Adult neurogenesis*

In the mature brain, neural stem cells mainly restricted to the subventricular and subgranular zones continue to generate new neurons throughout life. As these cells divide and differentiate, there is a unique window of opportunity for retrotransposons to mobilize. Factors that increase the rate of neurogenesis may likewise promote retrotransposition as a secondary effect. Exercise and exposure to complex or "enriched" environments are two factors known to drive plasticity in the brain. Exercise is believed to increase the number of newborn cells, whereas environmental enrichment enhances survival of those cells. When rodents are provided with a running wheel, the level of L1 retrotransposition in cells arising from the subgranular zone in the hippocampus, as measured by an L1-EGFP reporter construct, rises 3-fold compared to the level in sedentary rodents.[44] As these adult-born neurons in the hippocampus are believed to play a role in learning and memory, this finding raises questions around how retrotransposon-derived mosaicism in these cells could contribute to cognition. Studies in flies can begin to point toward an answer to these questions. In Drosophila, memory-relevant cells known as mushroom body neurons express approximately 2–4 times higher levels of transposons compared to other cell types. They are also deficient for piRNA proteins responsible for suppressing transposons. mushroom body neurons accumulate transposon insertions, with about half of them being near annotated genes and enriched in gene ontology terms related to neural function, including genes involved in mushroom body development and function.[45] Taken together, these results suggest that retrotransposon activity is an important source of brain plasticity in response to experience.

3.3 *Mature neurons*

Despite our knowledge of the prevalence of somatic retrotransposition within dividing NPCs, there remains a debate in the field as to whether mature post-mitotic neurons are capable of supporting retrotransposition. Given the ongoing pursuit to identify the downstream consequences of somatic retrotransposition, it is important to understand the cell types that can support this form of mutagenesis. For example, if somatic retrotransposition requires cell division, then it has the potential to encode events that are temporally restricted to the time point when the neuron was developing. Conversely, if post-mitotic neurons are capable of supporting retrotransposition, then these events have the potential to encode events throughout the lifetime of the cell, ranging from development of events triggered by neuronal activity to potentially deleterious events such as neuroinflammation. It is therefore important to determine whether post-mitotic neurons are capable of retrotransposition and, if so, the extent to which they support this process.

With respect to the case against post-mitotic retrotransposition, transposable element (TE) insertion events are thought to be restricted to dividing cells, largely due to the finding that L1 retrotransposition is inhibited by cell cycle arrest and replicative senescence.[34,46] Furthermore, the timing of L1-mediated retrotransposition is largely dictated by the phases of the cell cycle.[47] The nuclear membrane breaks down during mitosis, providing access to the large L1 machinery, which then remains in the nucleus through S-phase, and is required for successful completion of retrotransposition.[47] These studies suggest that L1 primarily enters the nucleus and facilitates retrotransposition as a function of the cell cycle. This does not preclude the possibility that somatic retrotransposition is still feasible, albeit rare, in post-mitotic neurons.

There is evidence to support the claim that L1 elements can integrate in non-dividing cells. While cell cycle studies suggest that L1 RNPs primarily enter the nucleus passively during membrane breakdown, there is evidence of active import of the L1 RNP.[48] L1-encoded Orf1p contains a non-canonical nuclear localization signal that associates with importin alpha, facilitating nuclear localization.[48,49] Furthermore, post-mitotic neurons *in vitro* have been shown to support low levels of integration of L1 elements.[50] Together, these studies indicate that, while nuclear membrane breakdown during the cell cycle is likely the primary method of L1 entry into the nucleus, it may still be possible for the L1 RNP to enter at low levels through active transport. Further work in the field is required to follow up on these results and validate that somatic retrotransposition does indeed occur in post-mitotic neurons *in vivo*.

4. Retrotransposition and Human Disease

Identifying the contribution of retrotransposition to disease is an important area of research, accelerated in recent years by advances in genome sequencing technology that have facilitated detection of *de novo* retrotransposon insertions. To date, there have been over 130 documented cases of human disease caused by retrotransposon insertions, though there are probably many more cases that have gone unidentified.[51] De *novo* insertions are predicted to occur in as many as 1 of 18.4–26.0 births, for example, in the case of Alu, but are not routinely discovered during clinical genome sequencing.[52] Recent analysis of retrotransposition events from over 9,000 whole-exome sequences produced by the Deciphering Developmental Disorders (DDD) study revealed that likely pathogenic insertions make up about 0.4% of diagnoses.[53] Though the rate of disease-causing germline retrotransposon insertions is relatively low in genetic disorders, the observation that there is a relatively high rate of somatic L1 retrotransposition in the brain has generated significant interest in studying the contribution of somatic insertions to brain-specific disease (reviewed in[54]).

4.1 *Schizophrenia*

Schizophrenia is a severe psychiatric disease that affects up to 1% of the population. Though the disease has a heritable component (80–85%), early environmental factors also play a reproducible role in the contribution to disease manifestation.[55] For example, prenatal infection is an environmental risk factor for schizophrenia, but its precise relationship to molecular and genetic determinants of disease is unclear.[56]

In 2014, Bundo and colleagues reported increased L1 retrotransposition in response to maternal immune activation using the poly-I:C model.[37] As mentioned above, Poly-I:C mimics viral double-stranded RNA and induces an immune response in animals. When administered to a pregnant mouse, the offspring show behavioral alterations reminiscent of schizophrenia, such as impairment in prepulse inhibition (i.e., a startle response test used in patients as well as mouse models) and social interaction.[57] A single injection of poly-I:C to mice during pregnancy was sufficient to increase L1 copy number in the prefrontal cortex of offspring, suggesting the rate of retrotransposition was increased in response to immune activation.[37] Further, neurons isolated from post-mortem patient brain tissue or from patient-derived induced pluripotent stem (iPS) cells exhibit increased L1 copy number. Whole-genome sequencing revealed that L1 insertions were enriched in genes related to synaptic function and schizophrenia.[37] These results provided the first evidence implicating retrotransposition as a mediator of environmental and genetic interactions in schizophrenia. In 2017, Doyle and colleagues extended this work in a separate patient cohort.[58] Using targeted genome sequencing, they confirmed the presence of L1 insertions in genes implicated in schizophrenia, with substantial overlap in the specific genes and pathways affected.[58] The molecular mechanism by which the L1 copy number is increased in schizophrenia remains to be determined, but recent studies have shown hypomethylation of L1 sequences in blood derived from schizophrenia patients.[41,59] Hypomethylation of repetitive elements is believed to reflect global dysregulation of methylation, which could contribute to derepression of L1 elements in the brain.

4.2 *Autism and Rett syndrome*

DNA methylation is critical for developmental regulation of gene expression and repression of retrotransposons. More than 90% of methylated cytosines occur in retrotransposons, and methyl-CpG-binding protein 2 (MeCP2) is a key player in methylation-mediated silencing of L1 elements.[60] Patients with Rett syndrome, a severe neurodevelopmental disorder that is a member of the autism spectrum disorder, harbor mutations in MeCP2 that result in derepression of L1 retrotransposition.[61] Neural progenitor cells differentiated from patient-derived iPS cells support a higher rate of retrotransposition, which is rescued by MeCP2 complementation. Further, an increased L1 copy number is observed in postmortem patient brain tissue compared to controls, though no difference was observed in heart tissue,[61] suggesting there is tissue-specific regulation of L1 retrotransposition upon MeCP2 loss. In a mouse model of MeCP2 knockout, L1 retrotransposition in the brain was over 3.5-fold higher compared to controls, again in a tissue-specific manner. Specifically, the cerebellum, striatum, cortex, hippocampus, and olfactory bulb were highly affected.[61] There may be developmental or cell type-specific mechanisms that protect certain tissues from acquiring more L1 copies upon MeCP2 loss, but those mechanisms have yet to be specified. Most recently, targeted sequencing of cortical neurons from Rett patients revealed that clonal somatic L1 insertions are enriched in introns and present in sense orientation, where they have the potential to disrupt transcription.[62] The significance of increased L1 retrotransposition for the progression of Rett syndrome is unknown. Some symptoms of Rett syndrome can be improved by reactivating MeCP2 during early life in mouse models, but it is not clear whether L1 insertions have occurred by this time or not.[63] The timeline of L1 accumulation and the contribution of high numbers of somatic mutations in the brain of Rett patients should be explored. Beyond Rett syndrome, autism spectrum disorders in general have been associated with low MeCP2 expression.[64] Recent observations point to reduced binding of MeCP2 at L1 sequences and increased expression of L1 in autism cerebellum, making this an exciting area for future study.[65]

4.3 *Alzheimer's disease*

Aging may be a natural risk factor for aberrant retrotransposon activity, as breakdown of repressive mechanisms allows aged neurons to become more permissive to retrotransposon expression. Therefore, the contribution of retrotransposition to age-related neurodegenerative disease has become an area of focus. Alzheimer's disease is the most common neurodegenerative disease in the United States and a major cause of dementia. Postmortem patient brain tissue is characterized by Tau protein, which is a contributor to genomic instability via chromatin relaxation, abnormal transcriptional activation, and DNA double-strand breaks.[66,67] Thus, Tau neurotoxicity could contribute to a permissive environment for retrotransposition. Indeed, analysis of more than 600 human transcriptomes revealed that expression of transposable elements broadly correlates with Tau burden, including LINEs, SINEs, and ERVs.[68] The effect appears to be dependent on Tau protein, because a fly model expressing pathologic misfolded Tau protein displayed similar increases in retrotransposon expression.[68] A major question remaining from this work is whether increased expression causes increased mobilization of retrotransposons in Alzheimer's disease. One study attempted to address this question by employing copy number assays to detect L1 sequences in about 40 postmortem patient and control samples, but failed to detect any significant difference.[69] Methodological limitations warrant that this work be repeated with a larger sample size using more sensitive techniques, such as single-cell genome sequencing, but nevertheless aberrant expression of retrotransposons on its own may have important implications for disease pathology by contributing to genome instability.

5. Alternative Mechanisms of Retrotransposon Toxicity

Insertional mutagenesis is one mechanism by which retrotransposons may contribute to brain disease, but it is not the only mechanism. Indeed, an emerging source of brain toxicity in neurological disease is the accumulation of cytoplasmic extrachromosomal nucleic acids derived from retrotransposons. In healthy cells, cytosolic nucleic acids are cleared from the cell via several complementary pathways. The ability to recognize abnormal buildup of nucleic acids, particularly those of foreign origin, is a major part of a successful immune defense. When detected, the presence of foreign nucleic acids triggers an antiviral defense program coordinated by Type 1 interferons. But, identifying the source of cytosolic nucleic acids – whether foreign or self – is imperfect and a number of studies have identified the buildup of nucleic acids as a source of autoimmunity, particularly in cases where the mechanisms used to clear nucleic acid debris from the cell begin to fail.

5.1 *Autoimmunity*

One such example is Aicardi Goutieres syndrome (AGS), an autoimmune encephalopathy linked to intracellular accumulation of DNA. AGS typically becomes apparent in infancy and leads to progressive neurologic and motor dysfunction that often results in death during early childhood. AGS has several possible genetic causes, one of which is Trex1 deficiency. Trex1 is a DNA exonuclease responsible for clearing nucleic acids, where loss of function results in severe Type 1 interferon response, an immune response related to anti-viral activity. Interestingly, Trex1 metabolizes reverse-transcribed DNA originating from retrotransposons, pointing to endogenous retroviruses and other retrotransposons as contributors to the autoimmunity observed in AGS.[70] In Trex1-deficient neural cells, L1 is a major contributor to the buildup of cytoplasmic DNA. Affected neurons undergo apoptosis, whereas affected astrocytes secrete Type 1 interferon, thus perpetuating the autoimmune response.[71] Likewise, Samhd1, another genetic cause of AGS, is a potent inhibitor of retroviruses, such as HIV, that works by depleting dNTP levels in the cytosol. Interestingly, retrotransposons are an endogenous target of Samhd1, and AGS-mutant Samhd1 fails to repress L1 retrotransposition.[72] Samhd1 may target L1 through multiple mechanisms, including by promoting the formation of stress granules to sequester cytosolic L1 ribonucleoproteins.[73] Taken together, these studies show that the consequence of unchecked cytoplasmic accumulation of retrotransposons seems to be severe autoimmune response.

In the absence of genetic deficiency in pathways mediating clearance of cytoplasmic nucleic acids, there are natural processes that can contribute to similar effects. During cellular senescence, Trex1 levels markedly decline, L1 elements become derepressed, and a Type 1 interferon response is triggered in response to cytoplasmic L1 DNA. Somatic activation of retrotransposons with age is a conserved phenomenon, occurring in groups of organisms such as yeast and drosophila.[74] In mice, treatment with reverse transcriptase inhibitors reduces the interferon response, suggesting that these extrachromosomal retrotransposons are contributors to age-associated inflammation.[75] Another mechanism leading to cytoplasmic retroelement accumulation with age is the loss of Sirt6, a regulator of L1 packaging into repressive heterochromatin. During aging, Sirt6 levels decline and previously silenced L1 elements become transcriptionally activated.[76] Interestingly, Sirt6 is regarded as a mediator of longevity in mammals, where loss of Sirt6 induces a premature aging syndrome.[77] Sirt6 knockouts exhibit shortened life span, accumulation of cytoplasmic L1 DNA, and a strong Type 1 interferon response that is rescued by reverse transcriptase inhibitors.[78]

5.2 *DNA damage*

Even in the absence of retrotransposition, the activation and expression of repetitive elements can contribute to DNA strand breaks and cellular damage or death. As one example, TDP-43, is an RNA-binding protein that has numerous functions, including regulation of RNA splicing, RNA stability, and repression of HIV. It has been implicated in neurodegenerative diseases, such as amyotrophic lateral sclerosis (ALS) and frontotemporal lobar degeneration, where its abnormal function leads to the formation of pathologic cytosolic aggregates. Notably, TDP-43 interacts extensively with transcripts derived from transposable elements (i.e., SINE, LINE, ERV, and some DNA elements) in the healthy cell. In neurodegenerative disease, the association of TDP-43 with its targets is lost and transposable elements become derepressed, where they gain the opportunity to contribute to genomic instability and DNA damage.[79] In a Drosophila model of TDP-43 pathology, a fly LTR-containing retrotransposon called gypsy is strongly activated in glial cells, which causes cell toxicity and death due to DNA damage. Blocking gypsy expression and treatment with reverse transcriptase inhibitors were both effective means to rescue the response.[80] Most recently, human post-mortem brain samples have been examined for signs of TDP-43 and retrotransposon pathology. Cortical samples from 148 ALS patients could be stratified by molecular subtype, with about 20% showing strong evidence of retrotransposon expression. This same subset showed evidence of TDP-43 pathology, suggesting that targeting of retrotransposons could be a clinically meaningful course of therapy for a certain subset of ALS patients.[81]

In Parkinson's disease, mesencephalic dopaminergic neurons undergo progressive degeneration, which can be recapitulated in mouse models by heterozygous deletion of Engrailed-1, a homeoprotein transcription factor believed to protect against oxidative stress, along with its paralogue Engrailed-2. In dopaminergic neurons of the adult mouse brain, full-length L1 sequences are expressed and can lead to oxidative stress-induced DNA strand breaks, mediated at least partially by L1 endonuclease activity. Overexpression of Engrailed directly suppresses L1 activity and the associated oxidative stress and protects against neurodegeneration in mouse models of Parkinson's disease.[82] In fact, multiple manipulations to decrease L1 expression (i.e., Piwil1 expression, anti-ORF2p siRNA, Engrailed expression) all protect against oxidative stress, suggesting that aberrant retrotransposon expression could be a general contributor to cell stress and genomic instability in many physiologic contexts.

6. Moving Forward

Here, we have reviewed the various impacts of retrotransposons on the mammalian brain, many of them deleterious, but moving forward, it will be interesting to observe what functional roles retrotransposons have in the mammalian brain. In early development, local transcription of L1 is required for chromatin changes that regulate gene expression, in *cis*, and thereby drive early embryonic development.[83] The finding that L1 may help to regulate chromatin structure is echoed in other systems such as in telomere maintenance and in Barr body inactivation.[84] Given that L1 expression is heightened in the neural progenitor state, it will be intriguing to examine if it plays a regulatory role in NPCs. Therefore, the coming years of retrotransposon biology may not only further elucidate their role in neurological disease but also identify their functional significance in the healthy human brain.

References

1. Mobile DNA III. *Mobile DNA* III. 2015. doi:10.1128/9781555819217.
2. Griffiths DJ. Endogenous retroviruses in the human genome sequence. *Genome Biol* 2001. doi:10.1186/gb-2001-2-6-reviews1017.

3. Bramham CR, *et al.* The Arc of synaptic memory. *Exp Brain Res* 2010. doi:10.1007/s00221-009-1959-2.

4. Pastuzyn ED, *et al.* The neuronal gene Arc encodes a repurposed retrotransposon Gag protein that mediates intercellular RNA transfer. *Cell* 2018. doi:10.1016/j.cell.2017.12.024.

5. Ashley J, *et al.* Retrovirus-like Gag protein Arc1 binds RNA and traffics across synaptic boutons. *Cell* 2018. doi:10.1016/j.cell.2017.12.022.

6. Irie M, *et al.* Cognitive function related to the Sirh11/Zcchc16 gene acquired from an LTR retrotransposon in eutherians. *PLoS Genet* 2015. doi:10.1371/journal.pgen.1005521.

7. Brandt J, *et al.* Transposable elements as a source of genetic innovation: Expression and evolution of a family of retrotransposon-derived neogenes in mammals. *Gene* 2005. doi:10.1016/j.gene.2004.11.022.

8. Chikamori H, Ishida Y, Nakamura Y, Koyama Y, Shimada, S. Distinctive expression pattern of Peg10 in the mouse brain. *Eur J Anat* 2019.

9. Uhlén M, *et al.* Tissue-based map of the human proteome. Science (80-.) 2015. doi:10.1126/science.1260419.

10. Lux A, *et al.* Human retroviral gag- and gag-pol-like proteins interact with the transforming growth factor-α receptor activin receptor-like kinase 1. *J Biol Chem* 2005. doi:10.1074/jbc.M409197200.

11. Nagasaki K, *et al.* Leucine-zipper protein, LDOC1, inhibits NF-kB activation and sensitizes pancreatic cancer cells to apoptosis. *Int J Cancer* 2003. doi:10.1002/ijc.11122.

12. Brattås PL, *et al.* TRIM28 controls a gene regulatory network based on endogenous retroviruses in human neural progenitor cells. *Cell Rep* 2017. doi:10.1016/j.celrep.2016.12.010.

13. Perron H, *et al.* Molecular characteristics of human endogenous retrovirus type-W in schizophrenia and bipolar disorder. *Transl Psychiatry* 2012. doi:10.1038/tp.2012.125.

14. Balestrieri E, *et al.* Children with autism spectrum disorder and their mothers share abnormal expression of selected endogenous retroviruses families and cytokines. *Front Immunol* 2019. doi:10.3389/fimmu.2019.02244.

15. Balestrieri E, *et al.* Human endogenous retroviruses and ADHD. *World J Biol Psychiatry* 2014. doi:10.3109/15622975.2013.862345.

16. Tartaglione AM, *et al.* Early behavioral alterations and increased expression of endogenous retroviruses are inherited across generations in mice prenatally exposed to valproic acid. *Mol Neurobiol* 2019. doi:10.1007/s12035-018-1328-x.

17. Kalkkila JP, *et al.* Cloning and expression of short interspersed elements B1 and B2 in ischemic brain. *Eur J Neurosci* 2004. doi:10.1111/j.1460-9568.2004.03233.x.

18. Liu WM, Chu WM, Choudary PV, Schmid CW. Cell stress and translational inhibitors transiently increase the abundance of mammalian SINE transcripts. *Nucleic Acids Res* 1995. doi:10.1093/nar/23.10.1758.

19. Policarpi C, *et al.* Enhancer SINEs link Pol III to Pol II transcription in neurons. *Cell Rep* 2017. doi:10.1016/j.celrep.2017.11.019.

20. Crepaldi L, *et al.* Binding of TFIIIC to SINE elements controls the relocation of activity-dependent neuronal genes to transcription factories. PLoS Genet 2013. doi:10.1371/journal.pgen.1003699.

21. Iacoangeli A, Tiedge, H. Translational control at the synapse: role of RNA regulators. Trends *Biochem Sci* 2013. doi:10.1016/j.tibs.2012.11.001.

22. Häsler J, Strub K. Alu elements as regulators of gene expression. *Nucleic Acids Res* 2006. doi:10.1093/nar/gkl706.

23. Eom T, *et al.* Neuronal BC RNAs cooperate with eIF4B to mediate activity-dependent translational control. *J Cell Biol* 2014. doi:10.1083/jcb.201401005.

24. Iacoangeli A, Dosunmu A, Eom T, Stefanov DG, Tiedge, H. Regulatory BC1 RNA in cognitive control. *Learn Mem* 2017. doi:10.1101/lm.045427.117.

25. Briz V, *et al.* The non-coding RNA BC1 regulates experience-dependent structural plasticity and learning. *Nat Commun* 2017. doi:10.1038/s41467-017-00311-2.

26. Zalfa F, *et al.* The Fragile X syndrome protein FMRP associates with BC1 RNA and regulates the translation of specific mRNAs at synapses. *Cell* 2003. doi:10.1016/S0092-8674(03)00079-5.

27. Lacar B, *et al.* Nuclear RNA-seq of single neurons reveals molecular signatures of activation. *Nat Commun* 2016;7.

28. McConnell MJ, *et al.* Intersection of diverse neuronal genomes and neuropsychiatric disease: the Brain Somatic Mosaicism Network. Science (80-.) 2017. doi:10.1126/science.aal1641.

29. Bodea GO, McKelvey EGZ, Faulkner GJ. Retrotransposon-induced mosaicism in the neural genome. *R Soc Open Sci* 2018. doi:10.1098/rsob.180074.

30. Feusier J, *et al.* Pedigree-based estimation of human mobile element retrotransposition rates. *Genome* Res 2019;29.

31. Erwin JA, *et al.* L1-associated genomic regions are deleted in somatic cells of the healthy human brain. *Nat Neurosci* 2016. doi:10.1038/nn.4388.

32. Denli AM, *et al.* Primate-specific ORF0 contributes to retrotransposon-mediated diversity. *Cell* 2015. doi:10.1016/j.cell.2015.09.025.

33. Muotri AR, *et al.* Somatic mosaicism in neuronal precursor cells mediated by L1 retrotransposition. *Nature* 2005. doi:10.1038/nature03663.

34. Xie Y, *et al.* Cell division promotes efficient retrotransposition in a stable L1 reporter cell line. *Mob* DNA 2013. doi:10.1186/1759-8753-4-10.

35. Coufal NG, *et al.* L1 retrotransposition in human neural progenitor cells. *Nature* 2009. doi:10.1038/nature08248.

36. Karijolich J, Abernathy E, Glaunsinger BA. Infection-induced retrotransposon-derived noncoding RNAs enhance herpesviral gene expression via the NF- B pathway. *PLoS Pathog* 2015. doi:10.1371/journal.ppat.1005260.

37. Bundo M, *et al.* Increased L1 retrotransposition in the neuronal genome in schizophrenia. *Neuron* 2014. doi:10.1016/j.neuron.2013.10.053.

38. Bedrosian TA, Linker S, Gage FH. Environment-driven somatic mosaicism in brain disorders. *Genome Med* 2016;8.

39. Meaney MJ. Maternal care, gene expression, and the transmission of individual differences in stress reactivity across generations. *Annu Rev Neurosci* 2001. doi:10.1146/annurev.neuro.24.1.1161.

40. Bedrosian TA, Quayle C, Novaresi N, Gage FH. Early life experience drives structural variation of neural genomes in mice. *Science* (80-.) 2018. doi:10.1126/science.aah3378.

41. Misiak B, *et al.* Lower LINE-1 methylation in first-episode schizophrenia patients with the history of childhood trauma. *Epigenomics* 2015. doi:10.2217/epi.15.68.

42. Nätt D, Johansson I, Faresjö T, Ludvigsson J, Thorsell, A. High cortisol in 5-year-old children causes loss of DNA methylation in SINE retrotransposons: a possible role for ZNF263 in stress-related diseases. *Clin Epigenetics* 2015. doi:10.1186/s13148-015-0123-z.

43. Fontana C, *et al.* Early Intervention in preterm infants modulates LINE-1 promoter methylation and neurodevelopment. medRxiv 2019. doi:10.1101/19011874.

44. Muotri AR, Zhao C, Marchetto MCN, Gage FH. Environmental influence on L1 retrotransposons in the adult hippocampus. *Hippocampus* 2009. doi:10.1002/hipo.20564.

45. Perrat PN, *et al.* Transposition-driven genomic heterogeneity in the Drosophila brain. Science (80-.) 2013. doi:10.1126/science.1231965.
46. Shi X, Seluanov A, Gorbunova V. Cell divisions are required for L1 retrotransposition. *Mol Cell Biol* 2007. doi:10.1128/mcb.01888-06.
47. Mita P, *et al.* LINE-1 protein localization and functional dynamics during the cell cycle. *Elife* 2018. doi:10.7554/eLife.30058.
48. Freeman BT, Sokolowski M, Roy-Engel AM, Smither ME, Belancio VP. Identification of charged amino acids required for nuclear localization of human L1 ORF1 protein. *Mob DNA* 2019. doi:10.1186/s13100-019-0159-2.
49. Goodier JL, Cheung LE, Kazazian HH. Mapping the LINE1 ORF1 protein interactome reveals associated inhibitors of human retrotransposition. Nucleic Acids Res 2013. doi:10.1093/nar/gkt512.
50. Macia A, *et al.* Engineered LINE-1 retrotransposition in nondividing human neurons. *Genome Res* 2016.
51. Kazazian HH, Moran JV. Mobile DNA in health and disease. *N Engl J Med* 2017. doi:10.1056/NEJMra1510092.
52. Stewart C, *et al.* A comprehensive map of mobile element insertion polymorphisms in humans. *PLoS Genet* 2011. doi:10.1371/journal.pgen.1002236.
53. Gardner EJ, *et al.* Contribution of retrotransposition to developmental disorders. *Nat Commun* 2019. doi:10.1038/s41467-019-12520-y.
54. Faulkner GJ, Billon V. L1 retrotransposition in the soma: a field jumping ahead. *Mobile DNA* 2018;9.
55. Cardno AG, Gottesman II. Twin studies of schizophrenia: from bow-and-arrow concordances to star wars Mx and functional genomics. Am J Med Genet — Semin Med Genet 2000. doi:10.1002/(SICI)1096-8628(200021)97:1<12::AID-AJMG3>3.0.CO;2-U.
56. Brown AS. Prenatal infection as a risk factor for schizophrenia. *Schizophr Bull* 2006. doi:10.1093/schbul/sbj052.
57. Meyer U, Feldon J. To poly(I:C) or not to poly(I:C): advancing preclinical schizophrenia research through the use of prenatal immune activation models. *Neuropharmacology* 2012. doi:10.1016/j.neuropharm.2011.01.009.
58. Doyle GA, et al. Analysis of LINE-1 elements in DNA from postmortem brains of individuals with schizophrenia. Neuropsychopharmacology 2017. doi:10.1038/npp.2017.115.
59. Li S, *et al.* Hypomethylation of LINE-1 elements in schizophrenia and bipolar disorder. *J Psychiatr Res* 2018. doi:10.1016/j.jpsychires.2018.10.009.

60. Yu F, Zingler N, Schumann G, Strätling WH. Methyl-CpG-binding protein 2 represses LINE-1 expression and retrotransposition but not Alu transcription. *Nucleic Acids Res* 2001. doi:10.1093/nar/29.21.4493.
61. Muotri AR, *et al*. L1 retrotransposition in neurons is modulated by MeCP2. *Nature* 2010. doi:10.1038/nature09544.
62. Zhao B, *et al*. Somatic LINE-1 retrotransposition in cortical neurons and non-brain tissues of Rett patients and healthy individuals. *PLoS Genet* 2019. doi:10.1371/journal.pgen.1008043.
63. Guy J, Gan J, Selfridge J, Cobb S, Bird A. Reversal of neurological defects in a mouse model of Rett syndrome. *Science* (80-.) 2007. doi:10.1126/science.1138389.
64. Nagarajan RP, Hogart AR, Gwye Y, Martin MR, LaSalle JM. Reduced MeCP2 expression is frequent in autism frontal cortex and correlates with aberrant MECP2 promoter methylation. *Epigenetics* 2006. doi:10.4161/epi.1.4.3514.
65. Shpyleva S, Melnyk S, Pavliv O, Pogribny I, Jill James S. Overexpression of LINE-1 retrotransposons in autism brain. *Mol Neurobiol* 2018. doi:10.1007/s12035-017-0421-x.
66. Frost B, Hemberg M, Lewis J, Feany MB. Tau promotes neurodegeneration through global chromatin relaxation. Nat Neurosci 2014. doi:10.1038/nn.3639.
67. Khurana V, *et al*. A neuroprotective role for the DNA damage checkpoint in tauopathy. *Aging Cell* 2012. doi:10.1111/j.1474-9726.2011.00778.x.
68. Guo C, *et al*. Tau activates transposable elements in Alzheimer's disease. *Cell Rep* 2018. doi:10.1016/j.celrep.2018.05.004.
69. Protasova MS, *et al*. Quantitative analysis of L1-retrotransposons in Alzheimer's disease and aging. *Biochem* 2017. doi:10.1134/S0006297917080120.
70. Stetson DB, Ko JS, Heidmann T, Medzhitov R. Trex1 prevents cell-intrinsic initiation of autoimmunity. *Cell* 2008. doi:10.1016/j.cell.2008.06.032.
71. Thomas CA, *et al*. Modeling of TREX1-dependent autoimmune disease using human stem cells highlights L1 accumulation as a source of neuroinflammation. *Cell Stem Cell* 2017. doi:10.1016/j.stem.2017.07.009.
72. Zhao K, *et al*. Modulation of LINE-1 and Alu/SVA retrotransposition by Aicardi-Goutières syndrome-related SAMHD1. *Cell Rep* 2013. doi:10.1016/j.celrep.2013.08.019.

73. Hu S, *et al.* SAMHD1 inhibits LINE-1 retrotransposition by promoting stress granule formation. *PLoS Genet* 2015. doi:10.1371/journal.pgen.1005367.

74. Kreiling JA, *et al.* Contribution of retrotransposable elements to aging. In Human Retrotransposons in Health and *Disease.* 2017. doi:10.1007/978-3-319-48344-3_13.

75. De Cecco M, *et al.* L1 drives IFN in senescent cells and promotes age-associated inflammation. *Nature* 2019. doi:10.1038/s41586-018-0784-9.

76. Van Meter M, et al. SIRT6 represses LINE1 retrotransposons by ribosylating KAP1 but this repression fails with stress and age. *Nat Commun* 2014. doi:10.1038/ncomms6011.

77. Mostoslavsky R, *et al.* Genomic instability and aging-like phenotype in the absence of mammalian SIRT6. *Cell* 2006. doi:10.1016/j.cell.2005.11.044.

78. Simon M, *et al.* LINE1 derepression in aged wild-type and SIRT6-deficient mice drives inflammation. *Cell Metab* 2019. doi:10.1016/j.cmet.2019.02.014.

79. Li W, Jin Y, Prazak L, Hammell M, Dubnau J. Transposable elements in TDP-43-mediated neurodegenerative disorders. *PLoS One* 2012. doi:10.1371/journal.pone.0044099.

80. Krug L, *et al.* Retrotransposon activation contributes to neurodegeneration in a Drosophila TDP-43 model of ALS. *PLoS Genet* 2017. doi:10.1371/journal.pgen.1006635.

81. Tam OH, *et al.* Postmortem cortex samples identify distinct molecular subtypes of ALS: retrotransposon activation, oxidative stress, and activated glia. *Cell Rep* 2019. doi:10.1016/j.celrep.2019.09.066.

82. Blaudin de Thé F, *et al.* Engrailed homeoprotein blocks degeneration in adult dopaminergic neurons through LINE-1 repression. *EMBO J* 2018. doi:10.15252/embj.201797374.

83. Jachowicz JW, *et al.* LINE-1 activation after fertilization regulates global chromatin accessibility in the early mouse embryo. *Nat Genet* 2017. doi:10.1038/ng.3945.

84. Bailey JA, Carrel L, Chakravarti A, Eichler EE. Molecular evidence for a relationship between LINE-1 elements and X chromosome inactivation: the Lyon repeat hypothesis. *Proc Natl Acad Sci USA* 2000. doi:10.1073/pnas.97.12.6634.

https://doi.org/10.1142/9789811249228_0009

Chapter 9

LINE-1 Mobilization in Cancers: More the Rule than the Exception

Daniel Ardeljan* and Kathleen H. Burns[†]

1. Introduction

Transposable elements (TEs) comprise roughly half of the human genome sequence,[1] and analysis of repetitive sequences has posed unique challenges for short-read sequencing. With the advent of targeted sequencing strategies,[2–5] improvements in bioinformatic methodologies,[6–8] and with newer long-read sequencing technologies in active deployment,[9] analysis of the structure and function of the repetitive genome has become increasingly tractable with the promise of providing further insight into the function of these repetitive DNA elements.

Understanding the genetic consequences of TE dynamics in relation to human health and disease has been a major focus of efforts in the last decade. These TEs include endogenous retroviruses (ERVs) contributing to anti-tumor responses associated with DNA hypomethylating drugs[10] as well as polymorphisms of *Alu* short

*Department of Medicine, The Johns Hopkins University School of Medicine, Baltimore, MD, USA.
[†]Department of Pathology, Dana-Farber Cancer Institute, Boston, MA, USA.

interspersed elements (SINEs) impacting disease risk (reviewed in[11]). Long interspersed element 1 (LINE-1, L1), as the only autonomous protein-coding retrotransposon currently active in the human genome, has been a major focus due to its ongoing ability to generate new insertions. Retrotransposition-competent L1 sequences are ~6kb in length and encode two open reading frame (ORF) proteins, the ~40-kDa ORF1 protein (ORF1p) and ~150-kDa ORF2p. ORF1p is an RNA chaperone and ORF2p has both endonuclease and reverse transcriptase activities.[12–14] Together, these proteins package the L1 mRNA into a ribonucleoprotein (or are hijacked by parasitic *Alu* short interspersed [SINE] or SINE-VNTR-*Alu* [SVA] elements), and then generate retroelement insertions by target-primed reverse transcription (TPRT).[15]

Several groups have noted that L1 activity can cause human disease[16,17] and generate somatic heterogeneity in the developing central nervous system,[18,19] the gastrointestinal tract,[20] and a multitude of cancer types.[3,6–8,20–30] The present chapter focuses on the role of L1 in cancers, as many groups have explored its expression, the mutational burden it places on tumors, and more recently its impacts on cell proliferation. It is our hope that in reading this review, investigators focusing on cancer biology will consider a simple question: "Is there a role for L1 in the cancer I study?"

2. L1 Cancer Genetics

A useful framework for thinking through the possible effects L1 has on phenotype is to consider L1 alleles as both inherited structural variants that are polymorphic in populations and as TEs that can mutate the genome to generate either driver or passenger insertions and rearrangements.

A given L1 allele can be polymorphic in a population (i.e., an individual either has the insertion or does not), a *de novo* acquisition within an individual's germline (i.e., present in the person's germline but absent in both parents'), or a *de novo* acquisition within a somatic cell (i.e., such as in a cancer with L1 reactivation). L1 insertions are then of variable length, ranging anywhere from a few DNA bases up to being a full-length element with or without a 3′ transduced segment of DNA. The median size of L1 insertions in the genome is ~900 base pairs.[1] An important point regarding L1 alleles that are polymorphic in the human population is that individuals do not typically differ with respect to the size of an L1 insertion at a particular locus, but rather differ in whether they have an insertion or not. Polymorphic L1s can cause phenotypic effects by mechanisms well described for other structural variants, including but not limited to affecting chromatin methylation patterns, recruitment of transcription factors, disruption of enhancers or other preexisting features, and so on. This potential has been demonstrated with other polymorphic TEs as well. Polymorphic *Alu* repeats are enriched at disease-risk loci[31,32] and can exert phenotypic effects by promoting the use of alternate splice sites[33,34] or modulating transcriptional output.[32,35] Catalogues of polymorphic L1 loci in human populations continue to expand as more individuals are sequenced, and so this number will likely grow as long-read genome assemblies become increasingly available. Several of the more active L1 source elements in cancer have been previously described,[6,8] but as we continue to identify new elements segregating in populations, this list may expand as well.

As a genome mutagen, an L1 insertion is either incompetent or competent for retrotransposition. To be retrotransposition competent, an L1 should be full length and have no premature stop codons in its ORF1 and ORF2 sequences. The mechanism by which L1 retrotransposes is called TPRT. Briefly, the L1 ORF2p endonuclease nicks genomic DNA to expose a 3′ OH. The poly-A tail of the L1 mRNA that is associated with the ORF2p is then used to prime the reverse transcription reaction. Retrotransposition-competent L1 loci are polymorphic in humans[36]: Any given individual possesses 80–100 such loci, but only approximately half are shared between individuals.[37] Among retrotransposition-competent loci, some have been shown to be highly active in cancers (so-called "hot" elements) and some are less so.[38] Retrotransposition-competent L1s are normally silenced and therefore inactive in somatic cells, but can reactivate in cancers and generate new insertions with functional consequences. Multiple instances of colorectal cancer initiation upon L1-mediated

APC tumor suppressor gene disruption have been described.[27,39,40] An L1 insertion disrupting the *PTEN* tumor suppressor has also been shown via whole exome sequencing in a patient with endometrial carcinoma.[23] In some instances, L1s are reactivated in precancerous lesions.[41,42]

However, while it is clear that somatically acquired L1 insertions can "drive" tumor progression, the field is still working out how and how commonly. Much of the targeted sequencing work in cancer genomes has revealed that acquired L1 insertions are rarely found to disrupt coding genes, let alone tumor suppressors, and are typically far in distance from transcribed loci. Even if L1 insertions act as "drivers" uncommonly, we likely are missing many instances of this in typical cancer genome studies. Sequence features of L1 insertions are distinct from those generated by other mutagenic sources. Specifically, canonical L1 insertions are defined by the presence of a target site duplication of variable sequence length, the L1 itself (which may be some length of poly-A tail and a variable amount of the L1 sequence), and a breakpoint in the genome with a consensus sequence of 5'-TTTT/AA-3 '.[13,43] These sequence features are difficult to map with short-read sequencing technologies and thus most reports to date are likely underestimates of the true extent of retrotransposition in cancer genomes. Furthermore, L1 insertions may also be associated with complex structural rearrangements including indels, copy number variants, duplications, inversions, or translocations (Fig. 1), and thus pose the combined bioinformatics challenge of finding both L1-associated features with these additional variants.

It is likely that L1 insertions, particularly in a tumor acquiring a large number of them, can impact tumor biology. Insertions that occur adjacent to or within introns of coding genes can be associated with transcriptional changes of these genes compared to normal tissues[22] or matched tumors without insertions in these same genes.[23] L1 insertions can supply novel splice and poly-adenylation signals as well as impact the transcriptional and translational efficiency of their target loci[23,44,45]; insertions could also generate transcripts and proteins from the L1 antisense promoter and ORF0 protein.[46,47] There have been few studies to analyze the impact retrotransposition may have on *cis*

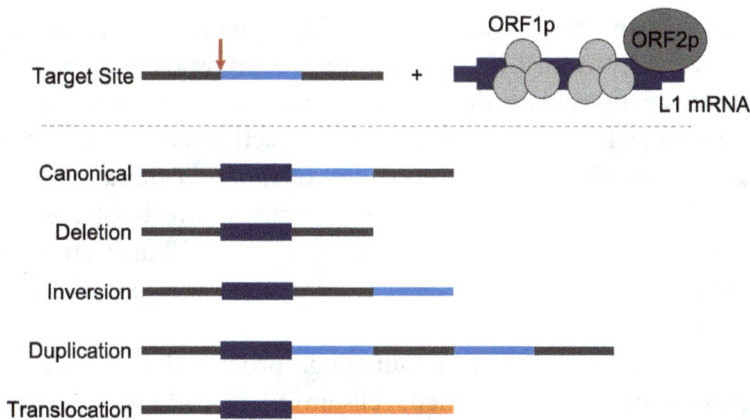

Figure 1. Variants associated with L1 insertions in cancers. Above the dotted line represents a genomic location with the red arrow indicating an L1 target site. Below represents the possible variants that may arise upon successful retrotransposition. Local L1 insertion sequence features include a target site duplication and the presence of a poly-A tail. Structural variation within the L1 includes internal inversion as a result of twin-priming and 5′ end truncations.[118] The canonical insertion would insert the L1 with target site duplication into the genome with no additional mutations. Other possible variants include deletions of genomic sequence, inversions, duplications, and translocations. These events may be responsible for oncogenic amplifications and/or tumor suppressor deletions.

regulatory elements. More recent work from the pan-cancer cohort has elegantly shown that L1 insertions can cause chromosomal-level structural changes (Fig. 1); repair of L1 insertions can induce template-switching events leading to duplications, inversions, and deletions, or even more catastrophic events such as bridge-fusion-breakage cycles resulting in either tumor suppressor loss or oncogene amplification.[8]

3. Reactivation of L1 in Cancer

Cancers that support L1 expression and retrotransposition include those derived from the endoderm: esophageal,[8,20,26] gastric,[8,25] hepatic,[3,30] pancreatic,[24] pulmonic,[6,8,23] colorectal[6,8,23,25,39,40]; genitourinary tissue such as breast,[6,8,23,24] prostate,[8] ovarian,[7,48,49] bladder[8]; and head and neck squamous cancers.[23] Retrotransposition in central nervous system cancers seems to be infrequent.[23,50,51] To date, no exhaustive analyzes of L1 expression or retrotransposition within histologically or molecularly defined tumor subtypes have been reported. Several groups have found evidence of L1 activation in precancerous ovarian lesions that correlate with loss of function mutations in the p53 tumor suppressor,[41,42] and p53 mutations have been correlated with L1 expression and retrotransposition.[8,23,29,52,53] Furthermore p53 has been shown to bind the L1 promoter and directly inhibit transcription.[54]

L1 retrotransposition is a mutagenic process that threatens the integrity of the genome. As such, cells limit L1 by several mechanisms that revolve around the life cycle of the element (Fig. 2). Some of these mechanisms are better understood than others. The possible stages of regulation are presumably the same as for any typical protein-coding gene locus: (i) transcription, (ii) mRNA processing and editing, (iii) protein translation, (iv) RNP formation and post-translational modifications, (v) DNA repair mechanisms at the time of an

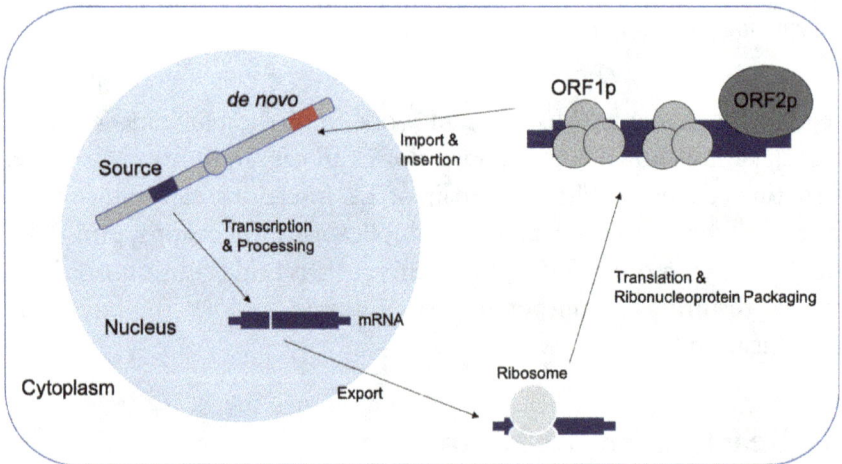

Figure 2. The L1 life cycle. A source L1 is transcribed and processed, then translated into ORF1p and ORF2p proteins and packaged into a ribonucleoprotein (RNP). The L1 RNP serves as a vector to transport L1 mRNA to a new genomic location and generate a de novo insertion. The chaperone activities of ORF1p and the endonuclease and reverse transcriptase activities of ORF2p are critical for this process. Each of these points in the life cycle presents an opportunity for the host cell to mitigate activity of L1s.

insertion, and (vi) as cells grow within a tissue. These regulatory mechanisms are variably affected in cancers, broadly speaking, which ultimately contributes to L1 activation and retrotransposition in many tumors.[52,55]

3.1 Transcription regulation

Retrotransposition begins with transcription of one of the relatively few "hot" L1 loci in the genome. The L1 5′ untranslated region (UTR) is a CpG-rich RNA polymerase II promoter[56,57] that interacts with a myriad of transcription factors.[58-63] The 5′UTR also promotes antisense transcription of ORF0 RNA; the ORF0 protein is not strictly required for retrotransposition but augments its efficiency.[46] The regulation of L1 transcription from individual loci has been shown to depend on the specific chromatin context (e.g., open chromatin epigenetic marks)[4] and methylation patterns of the promoter[27,30,41,49,64,65]; these same reports have demonstrated several instances in which loss of 5′UTR methylation is associated with L1 retrotransposition in human cancer tissues. As an example, Scott *et al.*[27] described in 2016 a "hot" polymorphic L1 that was hypomethylated in a patient with colorectal cancer, which allowed for the element's expression and subsequent insertion into the *APC* tumor suppressor. p53 has been shown to bind the L1 promoter and directly inhibit transcription, the consequence of which is to activate innate immune pathways at least in a melanoma cell line,[54] which thus raises the question regarding the interplay between L1 activity and tumor-associated immune responses. A genome-wide screen looking for transcriptional regulators identified the human silencing hub (HUSH) complex as a key mediator of L1 suppression[66]; the HUSH complex was originally described for its ability to target and silence lentiviral insertions[67,68] and also works on other transposon-mediated insertions such as sleeping beauty-mobilized transgenes.[69]

3.2 mRNA processing and editing

The vast majority of the L1 RNA sequence is generated from read-through transcription of insertions occupying introns or UTRs of protein-coding or non-coding genes.[70,71] Full-length "unit" or "authentic" L1 mRNA transcripts were first detected in a teratocarcinoma cell line[72,73]; these are approximately 6kb in length and span the 5'UTR, ORF1, a 63-base pair inter-ORF spacer sequence, ORF2, and a 3'UTR. They end in a poly-A tail that is required for retrotransposition due to its binding of ORF2p.[74] In some instances, splicing of the L1 RNA results in insertions with 5' truncations that are incapable of retrotransposition.[75] L1 transcripts can also fuse to the U6 small nuclear RNA to create chimeric transcripts.[76,77] These insertions are often unable to further retrotranspose, having 5'UTRs lacking transcriptional activators, generating transcripts with retrotransposition-incompetent ORF1p variants, or spliced ORF2p variants that can facilitate *Alu* but not L1 retrotransposition.[75,78–80] Splicing of mRNA is aberrantly regulated in a number of cancers and can result in tumorigenesis,[81,82] though the full extent to which splicing of L1 may be affected in cancers, both of retrotransposition-competent transcripts and transcripts containing read-through L1 sequence, should be explored further. Other described mechanisms of post-transcriptional regulation include uridylation of the L1 mRNA by terminal uridylase enzymes (TUT7 and TUT4), which inhibits the ability of L1 to reverse transcribe during TPRT,[83] and reductions in RNA stability via microRNAs such as miRNA-128.[84] Deaminase enzymes, in particular the APOBEC and ADAR proteins, have been shown to restrict L1 retrotransposition by mainly editing-independent (but some editing-dependent) mechanisms (reviewed in[85]).

3.3 *Translation*

Only a small fraction of full-length L1 insertions are capable of producing both ORF1p and ORF2p.[86,87] It appears that ORF2p translation is initiated after completion of ORF1p translation without L1 RNA dissociation from the ribosome[88]; this process remains obscure, as there is no obvious translation-initiation sequence in the inter-ORF sequence such as an internal ribosome entry site.[88] A previous study has explored L1 interactions with UPF1, a protein known for its role in promoting nonsense mediated decay (NMD) of transcripts with premature stop codons, and has demonstrated that these interactions are critical for promoting L1 retrotransposition while limiting protein production of L1 by an unclear mechanism seemingly distinct from canonical NMD.[89] More recent data have suggested that BRCA1 can directly limit the ORF2p protein amount via a protein–mRNA interaction in the cytoplasm.[90] There may be a role for the innate immune response, as it has been described that L1 expression induces type I interferons.[69,91]

Whereas ORF1p has been easy to detect in tumors, ORF2p has proven difficult. We recently developed high-quality ORF2p monoclonal antibodies that could detect ectopic expression of ORF2p from plasmids but not in cancer tissues, including some with confirmed somatically acquired L1 insertions.[87] This could signal that ORF2p expression levels are below the limit of detection by conventional assays (western blot, affinity purification with and without mass spectrometry) or that perhaps its expression is heterogeneous within tissue and with respect to time. ORF2p is a cytotoxic protein that causes replication stress,[69] and malignant populations likely are under selective pressures to reduce ORF2p expression.

3.4 *RNP formation and post-translational modifications*

Once successfully translated, retrotransposition-competent L1 organizes into a ribonucleoprotein (RNP) comprising ORF1p, ORF2p, and L1 mRNA.[92] These RNPs are abundant in the cytoplasm of cells overexpressing L1 from ectopic expression constructs and can organize into stress granules.[93,94] The RNP protein interactomes are diverse,[89,95,96] vary with the stage in the L1 life cycle purified (i.e., cytoplasmic vs. nuclear)[97], and include retrotransposition inhibitors such as the helicase MOV10 and antiviral protein ZAP.[96,98] Certain post-translational modifications such as phosphorylation of ORF1p have been shown to be required for retrotransposition,[99] though the exact mechanism and players involved remain to be defined.[100] Human (and mouse) ORF1p has been shown to be polyubiquitinated by the TEX19.1-UBR2 complex, which targets ORF1p for proteasomal degradation[101]; expression of this system is restricted to germline and pluripotent cells[102] and is activated by DNA hypomethylation during embryonic development, possibly in part to defend the germline from retrotransposition. Similar factors that may target L1 require further investigation.

3.5 *DNA repair of an insertion*

Arguably the most distinguishing aspect of L1 biology is the insertion mechanism, TPRT. The working model is that ORF2p nicks target site DNA with consensus sequence 5′-TTTT/AA-3′[13,43]; the L1 poly-A tail hybridizes to the thymidine track, and the exposed genomic 3′-OH is used to extend the daughter insertion by reverse transcription. This insertion begins from the poly-A tail and proceeds toward the 5′ end (Fig. 3). Most insertions are not full length. The completed insertion requires host factors to successfully incorporate into the genome, and typically consists of the L1 with poly-A tail flanked

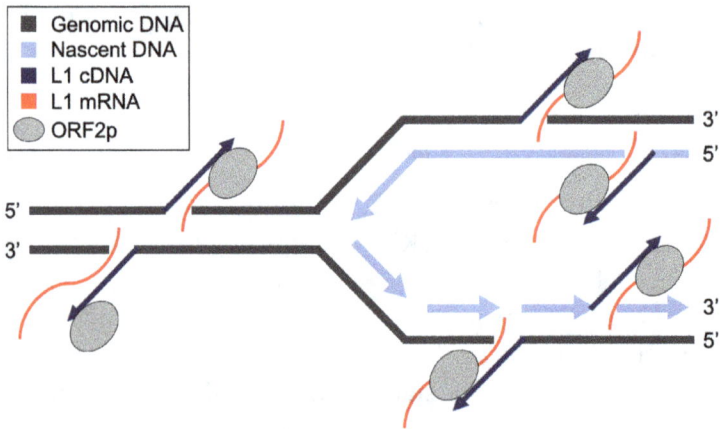

Figure 3. Possible orientations of L1 insertion attempts. Here, we illustrate possible insertion attempts into pre-replicated (to the left) and post-replicated (to the right) DNA. In all instances, L1 orientations are limited by whether the plus (+) or minus (−) strand is attacked based on canonical polymerase rules synthesizing DNA. Resolution or repair of insertion sites may depend on whether they occur in pre- or post-replicated DNA, and whether the insertions collide with replication forks causing fork stalling. A major hurdle to fully understanding how insertion intermediates are resolved is to identify the rules governing the second nick which completes the insertion.

by target site duplications (TSDs). While this is the canonical structure of a new L1 insertion, there are several variations observed both
in vitro as well as in vivo including 5′ inversions and target site deletions. Complex structural rearrangements are also possible including
L1-mediated deletions, translocations, tandem duplications, and
bridge-fusion-bridge cycles that can each affect megabases of DNA
sequence content.[8] The host factors that are critical in this process
include DNA repair pathways that recognize lesions, activate signaling cascades, and the repair machinery. We and others have demonstrated that retrotransposition is highest in S phase of the cell cycle
and incites replication-coupled DNA damage responses[69,103]; notably,
otherwise "normal" cells experience a p53-p21-dependent cell cycle
arrest in response to L1 expression.[69] These data, when interpreted in
the context of a complex tissue, would suggest that the response of
an individual cell to reactivation of L1 would be to cease dividing,
effectively inhibiting retrotransposition by limiting access to the S
phase of the cell cycle, thus halting the further generation of L1 insertions and limiting the production of somatic mosaicism at the tissue
level. A previous report found that aging results in L1 activation and
triggers an interferon response, which would induce senescence in the
affected cells.[91]

In a cancer cell that has evaded p53 signaling events, the cell cycle
proceeds and retrotransposition conflicts with DNA replication.[69] In
these cells, timing is of the essence. Insertions that occur in unrepli-
cated DNA are possibly repaired differently than insertions occurring
in post-replicated DNA or within replication bubbles (Fig. 3).
In vitro-generated insertions occur preferentially at endonuclease-
consensus target sites, are relatively depleted from transcribed genes,
and largely blind to local chromatin state; leading- and lagging-strand
bias has been observed in a cell-type-dependent manner and is
affected by EN-inactivating mutations.[104,105] In contrast, *in vivo* data
from human cancers has revealed that *de novo* insertions typically
occur in late-replicating chromatin and with a preference for open
chromatin.[8] The difference may lie in an insertion's interaction with
the replisome and replication-coupled DNA repair factors (Fig. 3).
Replicated DNA can be repaired by multiple mechanisms, affected by

the availability of a sister chromatid and the opportunity for homology-directed repair. In contrast, pre-replicated DNA only has its homolog available, which would differ in sequence and be separated in space. Many insertion intermediates might be excised and never leave their mark in the genome, and our work and others' have indicated a role for the Fanconi anemia complex in the repair of L1 lesions — which limits both DNA damage and L1 retrotransposition.[66,69,90] These considerations may suggest that the same mechanisms allowing replication machinery to access chromatin for the purposes of DNA synthesis expose a critical time window during which L1 can attack, much as any heavily defended fortress might be susceptible during a changing of the guard.

Ultimately, we have much to learn about the mechanisms governing successful completion of an insertion or repair of an insertion intermediate. Recent efforts to use forward genetic screens to identify genes impacting retrotransposition efficiency and cell fitness[69,90] have yielded useful insights. However, none of these studies have been carried forward to translational research models of cancer biology, and so how L1-associated DNA lesions are resolved in different diseases is not well understood. It is well known that DNA repair genes have epistatic effects, and mutations in some pathways can completely rescue perturbations to others.[106] Thus, further definition of retrotransposition modifiers and genetic dependencies of L1-expressing cells is needed in a multitude of cancer cell types. Inquiries into these parameters will be key to understanding the impact of retrotransposition in cancer and to identifying avenues to exploit this biology for therapeutic potential.

3.6 *Immune activation within a tissue*

Another potentially important question concerns how L1 activity impacts the tumor immune microenvironment. We and others have demonstrated that expression of L1 can induce an interferon response, which in some contexts prompts cell senescence and cell aging phenotypes, but in malignant contexts may have implications for tumor immunology.[69,91,107–109] Paradoxically, transcriptomic analysis of tumors stratified by number of somatically acquired L1 insertion has uncovered an anti-correlation between insertion rate and expression of immune-related genes.[29] If these are causally related, possible mechanisms include (1) immune signaling that directly limits key steps in L1 retrotransposition and (2) immune signaling that limits proliferation of L1-expressing cells that would be prone to retrotransposition, such that insertions would be possible, but these cells are less likely to clonally expand. Testing these hypotheses will require detailed studies of L1 biology in state-of-the-art organoid and *in vivo* cancer model systems and correlative studies in human tissues.

4. Conclusion: Assessing for Translational Impact of L1 Activity in Cancer

Our work has underscored the expression and activity of L1 in cancers and has motivated us to consider the potential of targeted therapies for L1-expressing malignancies. The appeal of targeted therapies is the promise of treating a cancer while limiting the toxicity experienced by patients. Perhaps the best-regarded example is the development of tyrosine kinase inhibitors used in chronic myeloid leukemia, and resulting in a relatively indolent disease course, significantly prolonging the life of patients, and featuring few side effects that disrupt patients' quality of life. Our interest in the possibility of a similar approach for L1-expressing tumors was inspired by the observation that despite the documented cytotoxicity associated with L1 activity, [110–114] tumors seem to tolerate this activity quite robustly. We recognized several years ago that the mere fact that cancers seem to thrive in the face of reactivation of this cytotoxic retrotransposon indicated that cancers may depend on specific mechanisms to limit this toxicity.

Recent work by us and others has pointed to a critical interaction occurring between retrotransposition and DNA replication. This finding suggests a model of L1 toxicity wherein L1 insertions create branched DNA structures that are positioned to interfere with forward progression of DNA replication forks during the S phase, which the cells are able to identify and repair by means of the replication-coupled stress response pathway including the Fanconi anemia machinery. Much remains to be learned, perhaps most importantly, how the genetic background and cellular context of a tumor might interact with L1 activity and affect possible treatment sensitivities. We speculate that amplifying L1 activity or undercutting cellular defense pathways might prove therapeutic in tumors supporting L1 retrotransposition, possibly synergistic in combination with DNA-damaging agents, or alternatively with inhibitors of ATR [115] or WRN helicase, [116] or perhaps even inhibitors of the spliceosome machinery.[117] Research over the last decade has clearly established that L1 is biologically active across many cancer types; among the most critical considerations now is whether this biology can be translated into clinical utility.

Acknowledgments

We thank Haig H. Kazazian, Jr. for his review of this chapter. The authors are supported by F30 CA221175 (D.A.) and R01 GM130680 (K.H.B.). Johns Hopkins University has licensed L1 monoclonal antibodies targeting ORF1p to EMD Millipore (Burlington, MA) and ORF2p antibodies to Abcam (Cambridge, UK). D.A. receives royalties from ORF2p antibody sales and K.H.B. receives royalties from ORF1p and ORF2p antibody sales.

References

1. Lander ES, Linton LM, Birren B, *et al.* Initial sequencing and analysis of the human genome. *Nature* 2001;409:860–921.
2. Solyom S, Ewing AD, Rahrmann EP, *et al.* Extensive somatic L1 retrotransposition in colorectal tumors. *Genome Res* 2012;22:2328–38.
3. Shukla R, Upton KR, Munoz-Lopez M, *et al.* Endogenous retrotransposition activates oncogenic pathways in hepatocellular carcinoma. *Cell* 2013;153:101–11.
4. Philippe C, Vargas-Landin DB, Doucet AJ, *et al.* Activation of individual L1 retrotransposon instances is restricted to cell-type dependent permissive loci. *eLife* 2016;5.

5. Steranka JP, Tang Z, Grivainis M, *et al.* Transposon insertion profiling by sequencing (TIPseq) for mapping LINE-1 insertions in the human genome. *Mob DNA* 2019;10:8.

6. Tubio JMC, Li Y, Ju YS, *et al.* Mobile DNA in cancer. Extensive transduction of nonrepetitive DNA mediated by L1 retrotransposition in cancer genomes. *Science* 2014;345:1251343.

7. Tang Z, Steranka JP, Ma S, *et al.* Human transposon insertion profiling: analysis, visualization and identification of somatic LINE-1 insertions in ovarian cancer. *Proc Natl Acad Sci U S A* 2017;114: E733–40.

8. Rodriguez-Martin B, Alvarez EG, Baez-Ortega A, *et al.* Pan-cancer analysis of whole genomes identifies driver rearrangements promoted by LINE-1 retrotransposition. *Nat Genet* 2020.

9. Audano PA, Sulovari A, Graves-Lindsay TA, *et al.* Characterizing the major structural variant alleles of the human genome. *Cell* 2019; 176:663–75 e19.

10. Chiappinelli KB, Strissel PL, Desrichard A, *et al.* Inhibiting DNA methylation causes an interferon response in cancer via dsRNA including endogenous retroviruses. *Cell* 2015;162:974–86.

11. Payer LM, Burns KH. Transposable elements in human genetic disease. *Nat Rev Genet* 2019;20:760–72.

12. Mathias SL, Scott AF, Kazazian HH, Jr., Boeke JD, Gabriel A. Reverse transcriptase encoded by a human transposable element. *Science* 1991;254:1808–10.

13. Feng Q, Moran JV, Kazazian HH, Jr., Boeke JD. Human L1 retrotransposon encodes a conserved endonuclease required for retrotransposition. *Cell* 1996;87:905–16.

14. Hohjoh H, Singer MF. Cytoplasmic ribonucleoprotein complexes containing human LINE-1 protein and RNA. *EMBO J* 1996;15: 630–9.

15. Luan DD, Korman MH, Jakubczak JL, Eickbush TH. Reverse transcription of R2Bm RNA is primed by a nick at the chromosomal target site: a mechanism for non-LTR retrotransposition. *Cell* 1993;72: 595–605.

16. Hancks DC, Kazazian HH, Jr. Roles for retrotransposon insertions in human disease. *Mobile DNA* 2016;7:9.

17. Kazazian HH, Jr., Moran JV. Mobile DNA in health and disease. *N Engl J Med* 2017;377:361–70.

18. Evrony GD, Cai X, Lee E, *et al.* Single-neuron sequencing analysis of L1 retrotransposition and somatic mutation in the human brain. *Cell* 2012;151:483–96.

19. Upton KR, Gerhardt DJ, Jesuadian JS, *et al.* Ubiquitous L1 mosaicism in hippocampal neurons. Cell 2015;161:228–39.

20. Doucet-O'Hare TT, Sharma R, Rodic N, Anders RA, Burns KH, Kazazian HH, Jr. Somatically acquired LINE-1 insertions in normal esophagus undergo clonal expansion in esophageal squamous cell carcinoma. *Hum Mutat* 2016;37:942–54.

21. Iskow RC, McCabe MT, Mills RE, *et al.* Natural mutagenesis of human genomes by endogenous retrotransposons. *Cell* 2010;141: 1253–61.

22. Lee E, Iskow R, Yang L, *et al.* Landscape of somatic retrotransposition in human cancers. *Science* 2012;337:967–71.

23. Helman E, Lawrence MS, Stewart C, Sougnez C, Getz G, Meyerson M. Somatic retrotransposition in human cancer revealed by whole-genome and exome sequencing. *Genome Res* 2014;24:1053–63.

24. Rodic N, Steranka JP, Makohon-Moore A, *et al.* Retrotransposon insertions in the clonal evolution of pancreatic ductal adenocarcinoma. *Nat Med* 2015;21:1060–4.

25. Ewing AD, Gacita A, Wood LD, *et al.* Widespread somatic L1 retrotransposition occurs early during gastrointestinal cancer evolution. *Genome Res* 2015;25:1536–45.

26. Doucet-O'Hare TT, Rodic N, Sharma R, *et al.* LINE-1 expression and retrotransposition in Barrett's esophagus and esophageal carcinoma. *Proc Natl Acad Sci U S A* 2015;112:E4894–900.

27. Scott EC, Gardner EJ, Masood A, Chuang NT, Vertino PM, Devine SE. A hot L1 retrotransposon evades somatic repression and initiates human colorectal cancer. *Genome Res* 2016;26:745–55.

28. Burns KH. Transposable elements in cancer. *Nat Rev Cancer* 2017;17:415–24.

29. Jung H, Choi JK, Lee EA. Immune signatures correlate with L1 retrotransposition in gastrointestinal cancers. *Genome Res* 2018.

30. Schauer SN, Carreira PE, Shukla R, *et al.* L1 retrotransposition is a common feature of mammalian hepatocarcinogenesis. *Genome Res* 2018;28:639–53.

31. Payer LM, Steranka JP, Yang WR, *et al.* Structural variants caused by Alu insertions are associated with risks for many human diseases. *Proc Natl Acad Sci U S A* 2017;114:E3984–92.

32. Wang L, Norris ET, Jordan IK. Human retrotransposon insertion polymorphisms are associated with health and disease via gene regulatory phenotypes. *Front Microbiol* 2017;8:1418.

33. Payer LM, Steranka JP, Ardeljan D, *et al.* Alu insertion variants alter mRNA splicing. *Nucleic Acids Res* 2019;47:421–31.

34. Clayton EA, Rishishwar L, Huang TC, *et al.* An atlas of transposable element-derived alternative splicing in cancer. *Philos Trans R Soc Lond B Biol Sci* 2020;375:20190342.

35. (a) Wang L, Rishishwar L, Marino-Ramirez L, Jordan IK. Human population-specific gene expression and transcriptional network modification with polymorphic transposable elements. *Nucleic Acids* Res 2017;45:2318–28. (b) Payer LM, Steranka JP, Kryatova MS, Grillo G, Lupien M, Rocha PP, Burns KH. *Alu* insertion variants alter gene transcript levels. *Genome Res* 2021; Online ahead of print.

36. Beck CR, Collier P, Macfarlane C, *et al.* LINE-1 retrotransposition activity in human genomes. *Cell* 2010;141:1159–70.

37. Brouha B, Schustak J, Badge RM, *et al.* Hot L1s account for the bulk of retrotransposition in the human population. *Proc Natl Acad Sci USA* 2003;100:5280–5.

38. Tubio JM, Li Y, Ju YS, *et al.* Mobile DNA in cancer. Extensive transduction of nonrepetitive DNA mediated by L1 retrotransposition in cancer genomes. *Science* 2014;345:1251343.

39. Miki Y, Nishisho I, Horii A, *et al.* Disruption of the APC gene by a retrotransposal insertion of L1 sequence in a colon cancer. *Cancer Res* 1992;52:643–5.

40. Cajuso T, Sulo P, Tanskanen T, *et al.* Retrotransposon insertions can initiate colorectal cancer and are associated with poor survival. *Nat Commun* 2019;10:4022.

41. Pisanic TR, 2nd, Asaka S, Lin SF, *et al.* Long interspersed nuclear element 1 retrotransposons become deregulated during the development of ovarian cancer precursor lesions. *Am J Pathol* 2019;189:513–20.

42. Xia Z, Cochrane DR, Tessier-Cloutier B, *et al.* Expression of L1 retrotransposon open reading frame protein 1 in gynecologic cancers. *Hum Pathol* 2019;92:39–47.

43. Cost GJ, Feng Q, Jacquier A, Boeke JD. Human L1 element targetprimed reverse transcription in vitro. *EMBO J* 2002;21:5899–910.

44. Han JS, Szak ST, Boeke JD. Transcriptional disruption by the L1 retrotransposon and implications for mammalian transcriptomes. *Nature* 2004;429:268–74.

45. Belancio VP, Roy-Engel AM, Deininger P. The impact of multiple splice sites in human L1 elements. *Gene* 2008;411:38–45.

46. Denli AM, Narvaiza I, Kerman BE, *et al.* Primate-specific ORF0 contributes to retrotransposon-mediated diversity. *Cell* 2015;163:583–93.

47. Criscione SW, Theodosakis N, Micevic G, *et al.* Genome-wide characterization of human L1 antisense promoter-driven transcripts. *BMC Genomics* 2016;17:463.

48. Xia Z, Cochrane DR, Anglesio MS, *et al.* LINE-1 retrotransposon-mediated DNA transductions in endometriosis associated ovarian cancers. *Gynecol Oncol* 2017;147:642–7.

49. Nguyen THM, Carreira PE, Sanchez-Luque FJ, *et al.* L1 retrotransposon heterogeneity in ovarian tumor cell evolution. *Cell Rep* 2018;23:3730–40.

50. Achanta P, Steranka JP, Tang Z, *et al.* Somatic retrotransposition is infrequent in glioblastomas. *Mob DNA* 2016;7:22.

51. Carreira PE, Ewing AD, Li G, *et al.* Evidence for L1-associated DNA rearrangements and negligible L1 retrotransposition in glioblastoma multiforme. *Mobile DNA* 2016;7:21.

52. Rodic N, Sharma R, Sharma R, *et al.* Long interspersed element-1 protein expression is a hallmark of many human cancers. *Am J Pathol* 2014;184:1280–6.

53. Wylie A, Jones AE, D'Brot A, *et al.* p53 genes function to restrain mobile elements. *Genes Dev* 2016;30:64–77.

54. Tiwari B, Jones AE, Caillet CJ, Das S, Royer SK, Abrams JM. p53 directly represses human LINE1 transposons. *Genes Dev* 2020.

55. Ardeljan D, Taylor MS, Ting DT, Burns KH. The human long interspersed element-1 retrotransposon: an emerging biomarker of neoplasia. *Clin Chem* 2017;63:816–22.

56. Swergold GD. Identification, characterization, and cell specificity of a human LINE-1 promoter. *Mol Cell Biol* 1990;10:6718–29.

57. Minakami R, Kurose K, Etoh K, Furuhata Y, Hattori M, Sakaki Y. Identification of an internal cis-element essential for the human L1 transcription and a nuclear factor(s) binding to the element. *Nucleic Acids Res* 1992;20:3139–45.

58. Becker KG, Swergold GD, Ozato K, Thayer RE. Binding of the ubiquitous nuclear transcription factor YY1 to a cis regulatory sequence in the human LINE-1 transposable element. *Hum Mol Genet* 1993;2:1697–702.

59. Tchenio T, Casella JF, Heidmann T. Members of the SRY family regulate the human LINE retrotransposons. *Nucleic Acids Res* 2000;28:411–5.

60. Yang N, Zhang L, Zhang Y, Kazazian HH, Jr. An important role for RUNX3 in human L1 transcription and retrotransposition. *Nucleic Acids Res* 2003;31:4929–40.

61. Athanikar JN, Badge RM, Moran JV. A YY1-binding site is required for accurate human LINE-1 transcription initiation. *Nucleic Acids Res* 2004;32:3846–55.

62. Harris CR, Dewan A, Zupnick A, et al. p53 responsive elements in human retrotransposons. *Oncogene* 2009;28:3857–65.

63. Sun X, Wang X, Tang Z, et al. Transcription factor profiling reveals molecular choreography and key regulators of human retrotransposon expression. *Proc Natl Acad Sci U S A* 2018;115:E5526–35.

64. Sanchez-Luque FJ, Kempen MHC, Gerdes P, et al. LINE-1 evasion of epigenetic repression in humans. *Mol Cell* 2019;75:590–604.e12.

65. Salvador-Palomeque C, Sanchez-Luque FJ, Fortuna PRJ, et al. Dynamic methylation of an L1 transduction family during reprogramming and neurodifferentiation. *Mol Cell Biol* 2019;39.

66. Liu N, Lee CH, Swigut T, et al. Selective silencing of euchromatic L1s revealed by genome-wide screens for L1 regulators. *Nature* 2018; 553:228–32.

67. Tchasovnikarova IA, Timms RT, Matheson NJ, et al. GENE SILENCING. Epigenetic silencing by the HUSH complex mediates position-effect variegation in human cells. *Science* 2015;348:1481–5.

68. Robbez-Masson L, Tie CHC, Conde L, et al. The HUSH complex cooperates with TRIM28 to repress young retrotransposons and new genes. *Genome Res* 2018;28:836–45.

69. Ardeljan D, Steranka JP, Liu C, et al. Cell fitness screens reveal a conflict between LINE-1 retrotransposition and DNA replication. *Nat Struct Mol Biol* 2020;27:168–78.

70. Deininger P, Morales ME, White TB, et al. A comprehensive approach to expression of L1 loci. *Nucleic Acids Res* 2017;45:e31.

71. Yang WR, Ardeljan D, Pacyna CN, Payer LM, Burns KH. SQuIRE reveals locus-specific regulation of interspersed repeat expression. *Nucleic Acids Res* 2019;47:e27.

72. Skowronski J, Singer MF. Expression of a cytoplasmic LINE-1 transcript is regulated in a human teratocarcinoma cell line. *Proc Natl Acad Sci U S A* 1985;82:6050–4.

73. Skowronski J, Fanning TG, Singer MF. Unit-length line-1 transcripts in human teratocarcinoma cells. *Mol Cell Biol* 1988;8:1385–97.
74. Doucet AJ, Wilusz JE, Miyoshi T, Liu Y, Moran JV. A 3' poly(A) tract is required for LINE-1 retrotransposition. *Mol Cell* 2015;60:728–41.
75. Larson PA, Moldovan JB, Jasti N, Kidd JM, Beck CR, Moran JV. Spliced integrated retrotransposed element (SpIRE) formation in the human genome. *PLoS Biol* 2018;16:e2003067.
76. Moldovan JB, Wang Y, Shuman S, Mills RE, Moran JV. RNA ligation precedes the retrotransposition of U6/LINE-1 chimeric RNA. *Proc Natl Acad Sci U S A* 2019;116:20612–22.
77. Buzdin A, Ustyugova S, Gogvadze E, Vinogradova T, Lebedev Y, Sverdlov E. A new family of chimeric retrotranscripts formed by a full copy of U6 small nuclear RNA fused to the 3' terminus of l1. Genomics 2002;80:402–6.
78. Perepelitsa-Belancio V, Deininger P. RNA truncation by premature polyadenylation attenuates human mobile element activity. *Nat Genet* 2003;35:363–6.
79. Belancio VP, Hedges DJ, Deininger P. LINE-1 RNA splicing and influences on mammalian gene expression. *Nucleic Acids Res* 2006;34:1512–21.
80. Belancio VP, Roy-Engel AM, Pochampally RR, Deininger P. Somatic expression of LINE-1 elements in human tissues. *Nucleic Acids Res* 2010;38:3909–22.
81. Escobar-Hoyos L, Knorr K, Abdel-Wahab O. Aberrant RNA splicing in cancer. *Annu Rev Cancer Biol* 2019;3:167–85.
82. Wang E, Aifantis I. RNA splicing and cancer. *Trends Cancer.*
83. Warkocki Z, Krawczyk PS, Adamska D, Bijata K, Garcia-Perez JL, Dziembowski A. Uridylation by TUT4/7 restricts retrotransposition of human LINE-1s. *Cell* 2018;174:1537–48 e29.
84. Hamdorf M, Idica A, Zisoulis DG, *et al.* miR-128 represses L1 retrotransposition by binding directly to L1 RNA. *Nat Struct Mol Biol* 2015;22:824–31.
85. Orecchini E, Frassinelli L, Galardi S, Ciafre SA, Michienzi A. Post-transcriptional regulation of LINE-1 retrotransposition by AID/APOBEC and ADAR deaminases. *Chromosome Res* 2018;26:45–59.
86. Penzkofer T, Jager M, Figlerowicz M, *et al.* L1Base 2: more retrotransposition-active LINE-1s, more mammalian genomes. *Nucleic Acids Res* 2017;45:D68–73.

87. Ardeljan D, Wang X, Oghbaie M, *et al.* LINE-1 ORF2p expression is nearly imperceptible in human cancers. *Mob DNA* 2020;11:1.

88. Alisch RS, Garcia-Perez JL, Muotri AR, Gage FH, Moran JV. Unconventional translation of mammalian LINE-1 retrotransposons. *Genes Dev* 2006;20:210–24.

89. Taylor MS, LaCava J, Mita P, *et al.* Affinity proteomics reveals human host factors implicated in discrete stages of LINE-1 retrotransposition. *Cell* 2013;155:1034–48.

90. Mita P, Sun X, Fenyo D, *et al.* BRCA1 mediated homologous recombination and S Phase DNA repair pathways restrict LINE-1 retrotransposition in human cells. *Nat Struct Mol Biol* 2020.

91. De Cecco M, Ito T, Petrashen AP, *et al.* L1 drives IFN in senescent cells and promotes age-associated inflammation. *Nature* 2019;566: 73–8.

92. Kulpa DA, Moran JV. Cis-preferential LINE-1 reverse transcriptase activity in ribonucleoprotein particles. *Nat Struct Mol Biol* 2006;13: 655–60.

93. Goodier JL, Zhang L, Vetter MR, Kazazian HH, Jr. LINE-1 ORF1 protein localizes in stress granules with other RNA-binding proteins, including components of RNA interference RNA-induced silencing complex. *Mol Cell Biol* 2007;27:6469–83.

94. Doucet AJ, Hulme AE, Sahinovic E, *et al.* Characterization of LINE-1 ribonucleoprotein particles. *PLoS Genet* 2010;6.

95. Goodier JL, Cheung LE, Kazazian HH, Jr. Mapping the LINE1 ORF1 protein interactome reveals associated inhibitors of human retrotransposition. *Nucleic Acids Res* 2013;41:7401–19.

96. Moldovan JB, Moran JV. The zinc-finger antiviral protein ZAP inhibits LINE and Alu retrotransposition. *PLoS Genet* 2015;11:e1005121.

97. Taylor MS, Altukhov I, Molloy KR, *et al.* Dissection of affinity captured LINE-1 macromolecular complexes. *Elife* 2018;7.

98. Goodier JL, Pereira GC, Cheung LE, Rose RJ, Kazazian HH, Jr. The broad-spectrum antiviral protein ZAP restricts human retrotransposition. *PLoS Genet* 2015;11:e1005252.

99. Cook PR, Jones CE, Furano AV. Phosphorylation of ORF1p is required for L1 retrotransposition. *Proc Natl Acad Sci USA* 2015; 112:4298–303.

100. Furano AV, Cook PR. The challenge of ORF1p phosphorylation: effects on L1 activity and its host. *Mob Genet Elements* 2016;6:e1119927.

101. MacLennan M, Garcia-Canadas M, Reichmann J, *et al.* Mobilization of LINE-1 retrotransposons is restricted by Tex19.1 in mouse embryonic stem cells. *Elife* 2017;6.

102. Kuntz S, Kieffer E, Bianchetti L, Lamoureux N, Fuhrmann G, Viville S. Tex19, a mammalian-specific protein with a restricted expression in pluripotent stem cells and germ line. *Stem Cells* 2008;26:734–44.

103. Mita P, Wudzinska A, Sun X, *et al.* LINE-1 protein localization and functional dynamics during the cell cycle. *Elife* 2018;7.

104. Flasch DA, Macia A, Sanchez L, *et al.* Genome-wide de novo L1 retrotransposition connects endonuclease activity with replication. *Cell* 2019;177:837–51 e28.

105. Sultana T, van Essen D, Siol O, *et al.* The landscape of L1 retrotransposons in the human genome is shaped by pre-insertion sequence biases and post-insertion selection. *Mol Cell* 2019;74:555–70 e7.

106. Kottemann MC, Smogorzewska A. Fanconi anaemia and the repair of Watson and Crick DNA crosslinks. *Nature* 2013;493:356–63.

107. De Cecco M, Criscione SW, Peterson AL, Neretti N, Sedivy JM, Kreiling JA. Transposable elements become active and mobile in the genomes of aging mammalian somatic tissues. *Aging* 2013;5:867–83.

108. Van Meter M, Kashyap M, Rezazadeh S, *et al.* SIRT6 represses LINE1 retrotransposons by ribosylating KAP1 but this repression fails with stress and age. *Nat commun* 2014;5:5011.

109. Simon M, Van Meter M, Ablaeva J, *et al.* LINE1 derepression in aged wild-type and SIRT6-deficient mice drives inflammation. *Cell Metab* 2019;29:871–85 e5.

110. Haoudi A, Semmes OJ, Mason JM, Cannon RE. Retrotransposition-competent human LINE-1 induces apoptosis in cancer cells with intact p53. *J Biomed Biotechnol* 2004;2004:185–94.

111. Belgnaoui SM, Gosden RG, Semmes OJ, Haoudi A. Human LINE-1 retrotransposon induces DNA damage and apoptosis in cancer cells. *Cancer Cell Int* 2006;6:13.

112. Gasior SL, Wakeman TP, Xu B, Deininger PL. The human LINE-1 retrotransposon creates DNA double-strand breaks. *J Mol Biol* 2006;357:1383–93.

113. Wallace NA, Belancio VP, Deininger PL. L1 mobile element expression causes multiple types of toxicity. *Gene* 2008;419:75–81.

114. Kines KJ, Sokolowski M, deHaro DL, *et al.* The endonuclease domain of the LINE-1 ORF2 protein can tolerate multiple mutations. *Mobile DNA* 2016;7:8.

115. Lecona E, Fernandez-Capetillo O. Targeting ATR in cancer. *Nat Rev Cancer* 2018;18:586–95.

116. Chan EM, Shibue T, McFarland JM, *et al.* WRN helicase is a synthetic lethal target in microsatellite unstable cancers. *Nature* 2019;568:551–6.

117. Salton M, Misteli T. Small molecule modulators of pre-mRNA splicing in cancer therapy. *Trends Mol Med* 2016;22:28–37.

118. Ostertag EM, Kazazian HH, Jr. Twin priming: a proposed mechanism for the creation of inversions in L1 retrotransposition. *Genome Res* 2001;11:2059–65.

Index

www.ingramcontent.com/pod-product-compliance
Lightning Source LLC
Chambersburg PA
CBHW050553190326
41458CB00007B/2018